Thinking Through the Box

Innovation Tools, Myths and Management

(Edition 2e)

Jagmohan Singh

Chief Trainer, **Friday Trainers**

jagmohanshan@gmail.com

www.fridaytrainers.co.in

@ All Rights Reserved

Innovation is the specific instrument of entrepreneurship, the act that endows resources with a new capacity to create wealth. - Peter Drucker

In Israel, a land lacking in natural resources, we learned to appreciate our greatest national advantage: our minds. Through creativity and innovation, we transformed barren deserts into flourishing fields and pioneered new frontiers in science and technology. - Shimon Peres

Out-innovating them is the way to beat China. - Jack Welch

One of the symptoms of an absence of innovation is the fact that you lose your jobs. Everyone else catches up with you. They can do what you do better than you or cheaper than you. And in a multinational corporate-free market enterprise, it is the company's obligation to take the factory to a place where they can make it more cheaply. - Neil deGrasse Tyson

In an era of endless innovation and constant disruption, what is any company really worth? How does a start-up determine its valuation? - Jay Samit

"Finding opportunity is a matter of believing it's there."
—Barbara Corcoran

"If at first the idea is not absurd, then there is no hope for it." —Albert Einstein

Dedicated to

Dr. Edward de Bono

Index

=====

The Cardinal Rule of Innovation
Prologue

Part A

Understanding Innovation   p13

i. Introduction to Innovation
ii. Innovation and Invention
iii. 4 Types of Innovation
iv. 4 Stages of Disruptive Innovation
v. Innovation Process
vi. Why Innovate?
vii. Where all can Innovation be Applied?
viii. Frame of reference
ix. Before you "Innovate", do this first
x. When is Innovation "risky"?

Part B

Tools for Innovation   p38

1. Thinking-Through-The-Box
2. If you can ask Questions, you can be an Innovator
3. Alternatives and Possibilities
4. Lateral
5. Redefine, Reposition
6. Deep Thinking

7. Concept Extraction
8. Artificial Crisis
9. Random
10. Accidental, Incidental
11. Provocation
12. Assumptions
13. Rules
14. NCs
15. Kaizen
16. Quality
17. Problems, Feedback, Suggestions
18. Observations
19. Psycho/Functional, Gain/Pain, Constraint
20. Competition
21. Future Already Happened
22. Tactics
23. Convergence, Synergy
24. Cross Application of Principles
25. Analogy
26. Matrix
27. Data Analytics & Pareto Magic
28. The 10 Ideas Everyone Has!
29. What It Isn't?

Part C

Myths of Innovation   *p1*30

1. Innovation is Only for Product (& Service) Designing
2. Innovation Helps Only in Sales & Marketing
3. Innovation – Only When Required

4. Innovation is done only by R & D Department
5. Innovation is a One Time Affair
6. Innovation is Costly
7. Innovation is Disruptive
8. You Have to Be a Genius, It Is Not For Everyone
9. Only Customers are Interested in Innovation
10. Innovation is Futuristic, Luxury for Present
11. Innovation is Difficult, Takes Time
12. Innovation Is All About A Dashing, Flashy Big Idea
13. Too Fancy A Thing To Be Of Interest To Management
14. Innovation has Cascading Effects / Implications
15. Innovation Creates New Problems
16. Innovation Happens Best Under Severe Constraints
17. Innovation Needs to be Managed
18. Least Risky Innovation is the Best
19. Innovation Guarantees Breakthrough in Business Success
20. Innovation is an Event
21. You have to Innovate yourself
22. You "do" Disruptive Innovation
23. Innovate on the Basis of Customer-Surveys
24. Disruptive Innovation needs Technological Breakthrough

Part D

Building Innovation-Capable Organization

*p*190

1. Innovation and Organization
2. Securing the Management
3. Company Wide Innovation Management
4. Innovation Training
5. Innovation Circles

6. Innovation Quantification, Targets, Monitoring & Showcasing
7. Resources
8. Environment, Culture and Ecosystem
9. Motivation, Appreciation, Rewards and Recognition

Epilogue *p222*

1. Wise Words from Innovation Masters
2. Facts and Figures

The Cardinal Rule of Innovation

"A new idea is delicate. It can be killed by a sneer or a yawn; it can be stabbed to death by a quip and worried to death by a frown on the right man's brow."

Charles Brower

Most of the new ideas that lead to breath-taking, game-changing, brilliant innovations are outright stupid ones at the first sight. So many destiny changing ideas have been assassinated much before they could be developed or recognised because of this natural human habit.

It must be ruthlessly ensured that how-so-ever strange, shocking, seemingly stupid, unintelligent, impractical, impossible, unaffordable, unwise, bizarre an idea, it must be welcomed with folded hands of the heart. It may pound your heart, but preserving and motivating it must be everyone's endeavour, more precious than their lives! The same applies for a very basic, simplest of an idea!

The merit of any Innovation lies in the honesty and zeal in the process of Innovation.

Prologue

The Innovative Solution

6 general managers of a company heading Finance, HR, Operations, Quality, Sales and R&D were taken to a remote resort by the Executive Director for a weekend brain storming workshop to find ways to reverse the sinking fortunes of the company.

They checked-in the resort on Saturday evening.

The director knew that he had to positively sort out the issues before the annual share-holders' meet. He had just two days to know the root cause of the slide despite the booming industry and flourishing competition!

Next morning, after a sumptuous breakfast, they all locked inside the room for the billion-dollar answer! The door opened twice, for lunch and supper but window to the solution remained tightly shut!

Early next morning the director asked for a cup of plain black coffee in his room! The waiter knocked at the door and went in with the cup of freshly brewed hot coffee!

"Can I have something for the headache!" the director asked the waiter!

"Surely, sir....... take a.........advice!" replied the bold waiter!

"What?" replied the visibly surprised director.

"Sack all the GM's except the R&D guy, sir. This will save your company"

The director was speechless! He remained silent for a few moments and then asked

"Why do you say that?"

"Last evening when I went to deliver his luggage, the HR guy didn't bother to thank me!

The Quality guy had thrown the cigarette buts in the lawn after smoking,

The Operations guy had his room all cluttered,

The Finance guy had left the ordered items in his lavish dinner unconsumed

And the Sales guy was using the rival brand!"

-

The director was shocked to hear all that! ...and from whom!

"And why should I spare the R&D General Manager?" the director curiously asked.

The waiter took a deep breath and said

"Because he has been asking me the recipe of every dish I am serving him!"

Part A

Understanding Innovation

i.

Introduction to Innovation

When you start with next to nothing, all you've got is a lot of thought, a lot of innovation, figuring new ways to do things without using a lot of money. - John Paul DeJoria

Suppose we are in the Innovation team of a health department and have been asked to come out with some innovative ideas about "Ambulance"!

That's right... "Ambulance"!

How will we do it? How will you do it?

Let's assume we call a team meeting, shut ourselves in a conference room, ask everyone to brainstorm and come out with ideas which we note down on a piece of paper and submit to the management or coordinator of the whole exercise.

Sounds easy, doesn't it?

Only that actual thinking and coming out with ideas is generally not that easy and simple. The ideas are slow to come and we all reach a dead-end sooner than later!

It's like banging ourselves like a bullet into the sandbags searching for ideas. We are not able to go too far into the sack of ideas before we are stopped!

Is there any other way to innovate? Is there any other way to approach this assignment?

Well, yes, there is!

In fact, not one, not two, but several...!

Let me first share some of the Innovative ideas about the Ambulances which I shall be randomly generating right now using the tools which I will be sharing in the subsequent pages of this book. (Caution: pl keep the cardinal rule of Innovation in mind. Also, we must not stop, discourage or play favourite or against any idea that comes in our mind.)

- How about having 2-wheeler and 3-wheeler ambulances to carry medicines and other essential equipments to patients instead of the other way around, besides having the ability to enter narrow roads and passages where wider 4 wheeler ambulances can't go?

- How about having paramedics or doctors drive these?

- How about ambulances which carry doctors instead of patients. (not moving the patients and instead taking the doctors to the patients in emergency situations)

- How about ambulances having a video-chat facility to let the doctors have a look at the patient and advise any emergency special advice or treatment as a stop gap measure?

- How about having flying ambulances (small helicopters) which can negotiate traffic jams wherever encountered?

- How about the use of drones to deliver medicines etc or to get and transmit the visuals and other details of an accident site?

- Ambulances stationed and scattered all over the cities, villages and highways etc instead of in hospitals, so that they can reach the patients anywhere within the shortest possible time

- Ambulances auto communicating with the traffic lights on the way to ensure that they are given priority safe passage at the crossings as they pass through

- How about fitting Ambulances with puncture proof or undeflatable tyres?

- Can we have more patients per (same) ambulance? How?

- Can we have mobile OTs and clinics and dispensaries on the lines of ambulances?

- Can we have linked sirens on the traffic signals in addition to the ones on the approaching ambulance to ensure open passages for it in advance?

- Can we have ambulance with doors on all three sides? With even roof removal if the need be?

- How can we have the critical tests and examinations started on the patient inside the moving ambulance?

- Can we have deboarding platforms at the hospitals which are at the height of that of the ambulances? Or ambulances with height adjustable and extendable platforms.

- Can we have mini ambulance units / compartments / corners in every bus truck train aeroplane?

- How about making it a standard procedure to ensure that the fuel tank of an ambulance is always filled to maximum after every trip?

- How about fitting GPS and other special software in ambulances to show the location of and connect with the nearest hospital or medical assistance available.

- How about having voice command operated interactive communication facility for the paramedic staff in the ambulance.

- How to make narrow width ambulances (for easier negotiating through traffic and narrow lanes) without sacrificing on the functionality.

- Having all Ambulances fitted with ABS, 4x4 and other safety features to enable it to remain safe during faster transportation.

- Equipping ambulances with biometrics linkage with data servers to know about patient's details and medical history on the move to understand.

- Designing and equipping ambulances in such a way that patient feels no jerk or swaying on the way.

These are a few random ideas that came to my mind right now, recalling some of the tools and techniques of Innovation shared in this book.

Several more can be generated as more brains join hands and spend more and focused time using the tools and techniques of Innovation.

How these and other ideas can be generated almost at will is all we shall be studying, witnessing and mastering in the subsequent pages.

ii.

Innovation and Invention

'Innovation' represents the process of 'Innovating'. The word 'Innovate' has origins in the Latin words 'in-' and 'novare' which respectively mean 'into' 'make new'.

While its more popular sibling 'Invention' means the creation of something new, innovation represents making 'new' of something which has already been created earlier.

That's really interesting and exciting definition! And needs a closer look...

So, while invention is the creation of something for the very first time, innovation is a sort of recreation of it albeit with a significant difference. It is like creating the same again (with clues from the existing one permitted liberally) in a different way (partially or wholly). While the thing essentially remains the same, it carries a sense of newness about it in whatever way!

While renovation (the third in the family of this word) is essentially just the restoration of the original shape and state of the invention, innovation is like renewing it in another way! A sort of value-added novelty in something already existing...

While God "invented" life, when confronted with "how many ways he could do so?" was "innovative" enough to create (several versions and species of) it repeatedly albeit in different ways each time!

Behind every innovation, there isn't merely artistic urge to do so, but also to add value, to improve it, to solve a problem, to come up with a different design, etc.

While an invention can't happen daily, innovation can and should, for the reasons discussed in detail at various places in the following pages.

Innovation essentially falls in two broad categories (the sub-categories of which are also discussed in next chapter) – Disruptive and Sustaining.

While Disruptive Innovation is the one which results in the replacement of an existing market and value network with a new market and new value network by (as the name suggests) disrupting the status quo (rather violently, drastically), Sustaining Innovation (whether evolutionary or revolutionary) is the one which improves the status quo without disrupting itself.

While Disrupting Innovation is brought about (so far) by the outsider, Sustaining Innovation is done by the entities themselves.

While the main purpose of Disruptive Innovation is an offensive strategy to capture the market and dethrone the incumbents, Sustaining Innovation is a defensive strategy by the incumbents to survive, improve and thrive.

> *Disruption, by our definition, means a shift in relative profitability from one prevailing business model to another. The dominant companies, accustomed to the old approach, lose market share to a new group of companies. Not every disruption is driven by advances in technology...competitors can emerge from seemingly anywhere. In sector after sector, new entrants are lowering prices, meeting consumer needs in novel ways, etc.*
> *- 10 Principles for Winning the Game of Digital Disruption (strategy-business.com)*

Having said that, though Disruptive Innovation, by sheer drama and excitement about it, is highly popular (and feared by every business, entity etc), it is only a part of the entire Innovation gamut.

I want to caution the readers from being swept away by just this one part and miss bulk of the other, and in turn becoming ever so more vulnerable and unhealthy.

iii.

4 types of Innovation

Radical (Revolutionary), Disruptive Innovation

= Popularly referred to as "disruptive innovation". Game changer...upsets the existing players as well as the market and systems dynamics. The market leader is often seriously challenged. Rules of the game change almost overnight. Highly pro customers... This marks the exploitation of a big gap in the value proposition to the customers or the introduction of an altogether new value paradigm which catches the imagination of the market. While this is the most dreaded and sought after form of innovation by the challengers (incumbents rarely opt for this) this is just one form of it. Though highly visible and dramatic, it doesn't replace the other three necessary forms of innovation which form the functional majority covering wide areas of applications often not covered under disruptive innovation.

Radical (Revolutionary), Sustaining Innovation

= This are big-leap innovations in any field or function which are giant leaps of improvements resulting in serious efficiency/productivity, design, process or application benefits solving some problem or resulting in some gain. These are the backbone of the whole innovation crusade and the bedrock of occasional revolutionary innovative ideas!

Non-Radical (Evolutionary), Sustaining Innovations

= These are the routine small time innovative improvements which are the prime indicators of the innovation culture in any organization. These are the tickers for a continuously improving healthy progressive company. Expect such companies to have tremendous inherent productivity and quality led competitive strength leading to a sustained bottom-line and playing saviour in difficult times.

Non-Radical (Evolutionary), Disruptive Innovations

= These are incidental/accidental, unsuspected, surprise, good-fortune routine innovations which disrupt the status quo without many drumbeats. These are essentially accidental/incidental triggering of trends or discovery of hidden/trapped potential and value. If triggered by a deliberate effort, mark the sign of ingenuity of the brain(s) behind it. These can as much be internal revolutions as well as external ones disrupting entire markets!

The top principle for disruptive and sustaining innovation is that it has to have a laser focus on customers. Innovation begins with their needs and expectations. Denise Morrison

iv.

Stages of Disruptive Innovation

Sudden abnormalcy in a comfortable trend

= Inevitable shocking disruption in a comfortable status quo in the ever-flowing stream of time. Disturbing but unavoidable. The earlier we acknowledge and understand the better for us in terms of facing and dealing with it.

Abnormalcy, uncomfortable trend

= Inevitable and natural post the above stage. Understanding leads to the enhancement in the pain and possible retaliation in a state of denial. The most difficult stage of the transformation process...the main zone of dismantling.

Normalcy in an uncomfortable trend

= Dust and smoke settling...normalcy returning to a still-unaccepted (though somewhat acknowledged) state of affairs...this is when the recovery starts taking the roots. This is where your response to the change starts taking shape. The reluctant start of reconstruction...

Normalcy in a comfortable trend

= Zone of reality acceptance, normalcy and adjustment to the new reality, paradigm, trend...

--

A disruptive innovation is a technologically simple innovation in the form of a product, service, or business model that takes root in a tier of the market that is unattractive to the established leaders in an industry. - Clayton M. Christensen

v.

Innovation Process

A dream will not become an innovation if there is no realization. - Ciputra

Innovation Processes can essentially be identified as four steps:

1. **Identification of Problem / Topic / Aim / Objective**

   = It is not advisable to wait for inspiration or for a drastic serious situation to emerge to opt for Innovation, it is of paramount importance that we find, learn and master ways to dig out or identify possible Innovation targets. While several may be roaming stealthily or openly around you, there may be others which may need to be force-created.

   In the words of Scott Berkun – "Discovering problems actually requires just as much creativity as discovering solutions. There are many ways to look at any problem, and realising a problem is often the first step toward a creative solution."

   Several Tools and Techniques have been shared in Part B of this book, with descriptions and examples. The list is largely indicative and not exhaustive. I keep on adding to the list as and when I stumble upon more (which I shall be adding in every subsequent edition of the book).

2. **Innovative Idea Generation**

= Once the challenge / problem / objective / aim / target / need for Innovation has been identified, the next stage if – actual "Innovation" – generating that big breakthrough idea.

For this too, several tools and techniques have been shared in part B. While the tools and techniques are meant to prompt, probe, inspire, instigate, nudge, indicate, suggest, and provoke, the hero is still the processor called the brain and the process called "burning of the midnight lamp"!

Sometimes the only kind of innovation comes when you have some solitude; when you step away. - Anand Mahindra

Often, the Innovation "artist" needs to be given absolute space and freedom during this phase and left undisturbed in their chosen environment!

If you look at history, innovation doesn't come just from giving people incentives; it comes from creating environments where their ideas can connect. - Steven Johnson

Intermittent team sessions can be arranged to recalibrate, shake-out, accelerate, adjust, recapitulate, re-align, synergize, etc.

3. **Idea Development**

    = Once the big idea (there can be more than one also) has been created, it is time to grow it, develop it, simulate it, understand the implications, test it...

    The Innovative Idea is like a magical spare part which needs to fit in a machine to deliver a full Innovation!

    This stage is a crucial one where key changes or adjustments need to be incorporated to save a potential gem of an Innovation from failing because of technical or oversight issues!

4. **Innovation Assimilation & Implementation**

    = By now, the Innovative idea has been identified, screened, chiselled and deburred, and is ready for deployment in the real war zone!

    This stage is akin to the actual transplant of a heart or kidney inside the recipient patient. Howsoever compatible the organ may be, if the recipient body rejects it, the patient can die.

    It has been seen that organizations are reasonably good at generating innovative ideas but the bottleneck is a lot further down the line.

    Similarly, howsoever brilliant an Innovation may be, its implementation is critical. It has to be as simple, natural, easy and seamless as possible. It may have G-force acceleration but without any jerk. We have to remember that the main purpose is value enhancement and not Innovation for the sake of it. It must not be forced. It has to be welcomed and assimilated by the receiver system as a much-awaited and much-needed "reform".

vi.

Why Innovate

to Stay in the Game

"Once household names have vanished or had their rock solid leadership eroded. Why? Because they stayed still, acknowledging the changing world around them, but refusing to act... or act quickly enough"... Shelly Greenway

to Stay Ahead of the Competition

Innovation distinguishes between a leader and a follower... Steve Jobs

to Find Affordable Solutions to Problems

One of the only ways to get out of a tight box is to invent your way out...Jeff Bezos

for Continuous Improvements

A sustaining innovation makes better products that you can sell for better profits to your best customers. Clayton Christensen

vii.

Where all can Innovation be applied?

Anywhere...

Everywhere...

for Everything...

viii.

Frame of Reference

Just like you need a handle to pick a cup or a sharp knife, you need a frame of reference to touch and pick the cup of challenges and the knives of problems. Frame of reference is to problems and concepts what point of reference is to space, time and motion.

You have to have a frame of reference to gauge, understand and deal with anything. Otherwise, it will be too slippery or incomprehensible a matter to handle or ponder about.

Point of reference helps us gain a fresh approach to the subject, looking at it with fresh pair of eyes. In fact, with fresh senses...

Barnes Wallis, in an interview, had said, "I knew nothing except how to think, how to grapple with a problem and then to go on grappling with it until you had solved it!"

People find innovation difficult because they don't know

- How to Think

- the Tricks, the Tools

This is exactly where the frame of reference pitches in.

Whenever you are stuck with a problem you are unable to understand or get the head or tail of it, you should try finding or creating a frame of reference.

The tools and techniques discussed in part A are nothing but ways and means to draw frames of references in different ways to try and unearth innovative breakthrough ideas.

Having the right frame of reference is equivalent to a quarter of victory in the Innovation battle.

ix.

Before You "Innovate" Ensure This First

"Innovation" is the latest storm in the buzz world! Everyone is talking about it. Everyone wants to innovate!

Ideally, nothing could be better news! Except that it is more froth than the stuff beneath!

I am often amused to hear this word from the industrialists, CEOs, managers, administrators besides others when I see them uttering this in the backdrop of where they stand, what and how they are managing!

Often, I see their customers and stakeholders, their infrastructures and processes, their cultures and ethics, their environments and policies, theirs vision and missions begging them to set the basics right.

I am aghast to see them turning deaf ear to their customers' users' and their own teams' loud requests, basic demands, feedbacks, complaints, suggestions, criticisms, murmurs, blabbers, expressions of disgust, and even abuses!

I am unable to understand their blindness to the countless ideas, points of attention and opportunities all around them and in their feet, waiting to be done to lift their companies and responsibility areas up to the basic hygiene levels!

I find them mouth big buzzwords in forums and business circles while hiding behind their B-suites and throat-controlling neck-ties while being mysteriously numb to the blatant disregard to the basics back in their "territories"!

In every company, organization, body, department, public place etc, customers after customers, users after users are directly indirectly, subtly loudly giving you feedbacks and suggestions which are nothing but clearest of clues about what all, if set right, could lift the customer satisfaction level and hence, the competitiveness of the organization dramatically and almost immediately.

It is shocking to see the leaders and companies aiming for greatness without first being good! Incomprehensible that we are wandering for the illusive difficult and dear options to escape the easy, economical and instant ones...!

There is little point in running after the lofty objectives of innovation if even the basic and open "pre-innovations" are not in place. Nothing could be a better example of being "penny wise, pound foolish" than this.

It is so easy to make things better, if only we are actually ready for it!

> Innovation is not a psychological therapy to feel good, but a holy path to be better when we are good.

x.

When is Innovation "risky"?

While solving a problem or breaking a deadlock, the innovation process is least risky simply because you are already in a spot and things can't get too much worse from where you already are. You have little to lose.

But when things are going fine, there are two scenarios - one, as a habit, you are endeavouring to keep improving things by constantly being on the lookout for innovative ideas. That, again, is not that risky because you are not trying to disrupt anything. You are only trying to accelerate or mould things gently, though firmly.

So, when is the "innovation" so called "risky"?

That's so, when you fear that some technology or development or new kid-on-the-block will disrupt you and you have not gathered courage to disrupt yourself voluntarily and pre-emptively.

It is not as much risky because of the risk involved but because of the fear of failure and lack of confidence.

What makes it doubly difficult for incumbent/defender companies is that they are no longer "young". Majority of them no longer have the energy, passion and risk-taking courage which the newer, younger and smaller challenger companies are brimming with!

This is precisely the reason why start-up skills never go out of fashion and reckoning. One should never grow so much as to lose the raw nerve and ability to start a new venture.

Every new addition, diversification or project by an existing company is like a new venture or start-up which faces similar challenges as regular new start-ups do. Therefore, the qualities required in the managing and promoting team are also similar. You can't manage a start-up just like you manage your grown up established company. Nobody owes you "another business". You would have to fight and survive for that as well.

Existing businesses survive the test of time by bearing own start-up-like versions while the parent business ages.

The risk to your business is not as much from the competition or innovation as from your diminished ability to fight like a start-up you once were.

This is exactly the advice Sunil Gavaskar gave Sachin Tendulkar to play larger innings. While Sachin, at the start of his career was finding it difficult to go sufficiently past 100 owing to lapse in concentration, Gavaskar advised him to take a fresh "guard" at the wicket post century and start a virtual fresh innings with a fresh mind!

Part B

Tools for Innovation

1

Thinking-Through-The-Box

Can you write an essay? Have you ever written one? Say, on your school, your granny, cow etc?

If you have, then there is great news for you!

You Can Innovate! You can be an Innovator! And a very good one! Innovation is as simple as that!

Don't believe me? Let me explain with an example...

Suppose we are asked to come out with some innovative ideas for...say, a shoe.

For that, all we have to do is write a very basic, short and simple essay (or a paragraph or so) on "shoe".

The idea is to explain the thing (here, a shoe) in a series of simple and very basic statements and observations, as if explaining it to someone who knows nothing about it and may have encountered it for the very first time. We have to explain in a way that even an alien would get a fair idea of the thing.

In the present example, let's assume that the listener we are explaining the shoe to, knows nothing about a shoe and has seen it for the very first time!

Let's try...

"We wear shoes on feet. Shoes come in pairs, one for each foot. Shoes are worn to protect the feet from dust and dirt, sharp objects on the road, and to preserve the soft skin from wearing out, for protection from temperature and weather, and so on.

Shoes are generally made of leather. Some are also made of cloth. Many shoes have laces on the top to tie so that they clasp the feet, remain fastened and don't go off while walking or running. Shoes are made by cobblers or in the factories. Shoes are bought from the shoe stores."

Now that is a pretty basic primary school essay which can fetch passing marks at the least...

But as I am about to show you, this simple and absolutely basic essay can be an extremely fertile base for several innovative ideas about the shoes.

I am not saying that the ideas I am going to extract from the above essay are going to be new and unique ones. This is just a random example and the idea is to demonstrate the technique.

What we need to do is examine/underline/mark/take out all keywords and phrases in the essay, and thereafter, to challenge them all with questions like "why?", "why not?", "what if?" "what else?" etc.

Some of the phrases/words easily visible in the essay are as follows (I am including the challenges to these in the brackets)

- **We wear shoes on feet.** (Why do we wear? Why can't shoe be worn without having to wear them? Can the shoes wear themselves around the feet without 'we' having to wear them? Can shoes be worn on parts other than feet (with changes, of-course)? Why do we have to wear the shoes? Is there any other way? Are such shoes possible which needn't be worn or taken off?)

I can see some of you are laughing while others are shaking their heads finding all this stupid and non-sense.

Well, I request you to re-read the Cardinal Rule of Innovation shared before this chapter.

Howsoever stupid the idea, it is not to be rejected, ignored or ridiculed. It has to be welcomed without judging.

(I will demonstrate how this seemingly simple technique can be used from genetics to aerospace engineering, social sciences to psychology, garments design to legal services, education to armed forces, health care to sports, etc.)

- **Shoes come in pairs**....Why? What else is possible? Are such shoes possible which can be worn on any foot - left or right? How about such a shoe design?

- **Shoes are worn to protect the feet**.....What else can shoes be used for? Worn for? To enhance performance? To reduce shock to the foot while jumping, landing, running etc? To apply medications? To measure body parameters? As data storage devices? As cameras? As power generators? To record walking and running performance?

- **Shoes are generally made of leather**...Why? What else is possible?

- **Shoes have laces**...Why? Can there be shoes without laces? What all alternatives can we have?

- **Laces are on the top of the shoes**...Where else can we have the laces if we are to have the laces only? Are self-tieing laces possible?

- **Shoes are made in factories or by cobblers**...Where and how else can shoes be made?

- **Shoes are sold in shoe-stores**...Where all can shoe be sold and purchased from? Can we purchase without going to stores?

The possibilities are endless, limited only by our imagination. Imagination only depends on the question we can ask. The questions are all having their hints in the words and phrases in the essay.

We can do the same for any object/product, phenomenon, concept, process, situation, dilemma, challenge, set-up, etc.

The idea is to spread open, dissect, unassemble, any concept, process etc in such a way that its every constituent part comes to forth for analysis and scrutiny.

Once that is done, all we have to do is hold every layer, part and constituent, and expose it to all sorts of open and closed ended questions, including those hitherto never thought and asked.

If you open up the mind, the opportunity to address both profits and social conditions are limitless. It's a process of innovation. - Jerry Greenfield

So, all you have to do to be an innovator is to write an essay, a simple 2nd or 3rd-standard essay, a thorough but basic one. Like a child...without hesitation...

Everything has its next innovation hidden inside it.

Thousands of years of improvements and innovations in the human history are nothing but a long chain of changes and breakthroughs in the already existing things and concepts revealed by observations triggered by questions.

"Why?", "Why not?", "Why not?", "What else?" etc are godly questions.

The boxes in the human history have been emerging whenever humans have dared or responded to the urge of thinking out of the box. But, it is conveniently forgotten that when we think out of the box, the box is still there as the critical reference point. It is this box which is leading us to outside the box. It is the visible yet invisible link to the unending chain of boxes of things and concepts.

I fondly call this process of innovation **"thinking-through-the-box"**! It works from genetic engineering to economics to product designing to service improvements to process re-engineering to governance challenges etc.

Everyone talks about thinking out of the box and in the process throws out or ignores the existing box, conveniently forgetting that the box is full of the universe of possibilities outside the box. The box itself is begging and desperate to take you on the adventure beyond it, through it.

-

Let's take another example.

Suppose, we are asked to come out with innovative ideas about an aeroplane!

Let's try to explain an aeroplane with basic statements. (Explaining every part, function, technology etc would take months and years to explain and a library of books and manuals; so what we are doing here is just a small demonstration of absolute basics. But the technique can be applied to every sub-assembly, part, system, technology etc of the aeroplane separately and one by one to come out with innovative ideas. In fact, that is why it is said that Innovation is a team work, and Innovation has layers under layers!)

- Aeroplane is a tubular flying vehicle which can carry passengers as well as cargo over land and sea.

- It has two main wings, one on either side, bent towards the rear, two small wings at the rear and a vertical tail. Both the main and the rear wings have got flaps which can be moved to navigate the direction of the plane.

While a lot can be written about the plane, let's see, as of now, what all ideas are peeping out of the two small paragraphs above.

- Can aeroplanes only be tubular? Can aeroplanes only "fly"? Can aeroplanes be used for any other purpose other than transporting people and cargo?

- Is it a compulsion for the aeroplane to have 2+2 wings? Can any other arrangement be done? Is tail of the aeroplane compulsory to have? Is navigation of the plane possible in some other manner as well?

Quite a few of these points might seem and sound crazy, stupid, childish, impossible and impractical. That is why it is advisable to keep a page close by with the cardinal rule of innovation written on it!

Almost all great innovations and inventions were considered stupid and impossible before they actually happened.

For Innovation we have to look - where we are not used, think - what we are not used to, challenge - what we are not habitual to!

To fast forward the example, I am sharing below some of the ideas popping in my head about the aeroplanes:-

- Is it a compulsion for the aeroplane to run a distance on the air strip to take off? Can there be any other way?

- Are tyres / wheels (on the landing gears) the only solution for a plane to stand, take-off and land on?

- Does the cockpit have to be there in the nose tip of the plane compulsorily?

- Do the pilots have to be there?

- Why does the aeroplane have one or two doors only? Why it has "doors"? What other way can be there to board / de-board the passengers and cargo?

- Do the turbines have to run of Aviation Jet Fuel only? What other options can be there?

- (To use a very aggressive tool that Dr. Edward de Bono calls "Po") Will the humanity cease to fly if turbines were not to be made or the day jet fuel supplies dry up? Can aeroplanes fly without turbines? With something else? What options do we have or can be considered?

- Why do aeroplanes have so small windows in the main fuselage?

- Why does the aeroplane have to be single decker? (already incorporated in Airbus A380)

- How high a plane can fly? Why? Why not? What are the possibilities?

- Why do the floor and the roof of the plane have to be opaque? Why can't the passengers see the sky or down below? Why? Why not?

- Why can't the passengers use telephony or internet in flight? (already incorporated)

- Why do passengers have to be confined to their seats all through the journey? What can be done?

- Can aeroplanes land only on air-strips? Where all can they land on? Can they land vertically straight down like Harrier fighter planes? Why not? How can they be done to make them capable of doing that?

- How really big can aeroplanes be? And how small? Why? Why not?

- How wide can aeroplanes be? And how tall? Why? Why not?

- Why do the planes have to compulsorily bear the turbulence? Why? Why not? What can be done?

- How can the bird hit incidents be eliminated? How can a bird hit be rendered harmless?

- How can a commercial airliner be safeguarded from the hostile missiles fired at it (like what happened in Ukraine to flight MH17). What can be done? What are the limitations? What can be done?

Once you get a good lot of ideas, you can settle down to segregate, scrutinize and study them thoroughly from fitment, practicality, feasibility and other relevant points of view.

There were times, television were fat and bulky. Questioning that opened the way for slimmer ones. Physical keypads on the mobiles were challenged, resulting in their vanishing act! Someone noticed the thick bezels of the smart phones and questioned them. Result? They are almost gone!

Somebody noticed and questioned the charging speed, resulting in a dramatic improvement. Somebody noticed and questioned the compulsion of wired charging and here we have the wireless charging!

Somebody challenged the compulsion of having driver inside a car and there we have the emerging scenario of a driverless car. Somebody challenged the Internal Combustion Engines and we have the electric cars.

Everything and everything we are noticing, stating and challenging is coming in the line of the sharpshooter called Innovators! And everything we are ignoring or refusing to see or acknowledge is there like a blessed curse!

Every box tells us what and where to look for, outside! The box wants to get out of itself...begging us to think out of it through its eyes!

We must learn to think through the box!

2

If you can ask Questions, you can be an Innovator

Companies have too many experts who block innovation. True innovation really comes from perpendicular thinking. - Peter Diamandis

The only advantage humans have got over all other species is the ability to think...astronomically more than even some of those animals which can "think"!

What a tragedy that thinking is what we have almost but stopped.

What we do these days in the name of thinking is mere pattern recognition and sub-conscious selection from past experiential data in our heads.

We are so overwhelmed by the state of affairs that we accept anything and everything which has sufficient web of magic to further overwhelm us.

We are living beautiful numb lives.

We have stopped questioning except when pushed to the limits.

We have lost the art of questioning.

We have forgotten the power of the questions.

Questions need curiosity and courage. You don't have to be a genius to ask questions. On the contrary, an ignorant curious person can ask more and better questions.

Now the question is - why ask questions?

Because not asking questions is a sign of acceptance of status-quo. Not asking questions is acceptance of all that is, where is, as is. Period. It marks the end of all progress. It denotes the comfort zone unknowingly but inevitably drifting towards decay.

The progress of the human race would have been only 1 or 2 % of what has been had we confined ourselves to mere accidental and incidental innovations and improvements. An overwhelming majority of the breakthroughs and innovations are the outcome of answers forced out because of innocent questions.

Therefore, ability to ask questions is an ability which straightaway puts you amidst the class of natural innovators.

What sort of questions? - You will ask.

Well, all those questions which come to your mind when you are not conscious, when you acknowledge all questions without getting ashamed of them...

Filtered, shortlisted intelligent questions are mostly useless, juiceless, dead, roughage ones which are not much productive.

Simple, honest, innocent, unashamed questions are the source of most stunning or effective innovations.

In the words of Dr.Edward de Bono, the father of lateral thinking, ask "searching questions!"

Why, What, How, Why not, What about, What if, What else, How about, etc are some of the searching questions which if posed even to the seemingly settled questions can yield answers which can open floodgates of innovations with corresponding cascading effects.

One more technique of questioning is – "Ask why 5 times".

The main purpose of this is to go to the root of the issue instead of merely bouncing from the superficial status.

For the Innovation to pass the test of Performance it must not be merely beautiful, but functional and effective. It must be an alien to the whole set-up but the invisible functional backbone.

The solution has to be an epitome of profound though invisible understanding of the purpose and constraints.

For that, we have to go to the depth the object and subject before attempting to innovate. A gorgeous innovation may be plain junk if not sensitive to the reality and dimensions.

Good innovators question anything and everything. They take nothing for granted. Everything is open to scrutiny for their eyes and minds.

Why do we do what we do? How we do, when we do, the way we do? Etc.

Answers are shy. They rarely raise their hands till the question is raised.

Questions are potential innovations. Those who can ask questions are potential innovators.

> **Challenge assumptions by asking an outrageous 'What if?' question. 'What if we could download the entire internet and analyse all the links.' It sounded impossible but they (google founders – Page and Brin) did it. – Paul Sloane**

3
Alternatives and Possibilities

Innovation requires an experimental mindset. - Denise Morrison

Once we have laid bare, spread the situation in a thin layer and exposed them to the ruthless and un-pardoning spotlight of all sorts of "searching" questions (as discussed in the previous two chapters), we are likely to find ourselves knocking at the doorsteps of the possible innovation.

By now, it starts getting amply clear that the situation, the status-quo of the thing under our focus is far from optimal or unchallengeable. The cracks start to appear.

The inefficiencies, the flaws, the shortcomings, the ugliness, the possible improvements, the gaps, the open-ends etc are all suddenly glaringly visible.

The purpose of the essay and questioning stages was to find out the points and directions of possible innovations.

It is like the broad identification of the options available.

Thereafter, all we have to do is search and list down all the alternatives and possibilities.

These include the semi or partial (even non-) possibilities as well.

After this stage, we have in our hands, the heap of unfinished raw contours of emerging innovations.

Though the final innovations may be dramatically different from what we have in our hands by now, all those will, inevitably be owing to the butterfly effect of this heap and this process.

***

4

Lateral

Not every breakthrough or innovation pops out on the immediate left, right or opposite side of the apparent place of it.

Many, if not most, innovations are the outcome of plain lateral thinking (a term coined and a concept immortalized by the legendary Dr. Edward de Bono).

For that, you have to think laterally...literally.

And literally, lateral means "from the side or sides"

-

Oxford English Dictionary defines 'lateral thinking' as 'seeking to solve problems by unorthodox or apparently unorthodox methods'.

We are generally so obsessed with looking at a thing or situation from one perspective that not only we become numb to the details, vulnerabilities, and possibilities from that angle we also forget to see other possible solutions from alternate views and approaches.

What we often call "lateral" is nothing but the ignored, unorthodox, non-traditional, non-habitual, uncomfortable, unusual way of looking at the subject.

Mindless habitual behaviour is the enemy of innovation. - Rosabeth Moss Kanter

We have been so used to imagining the gunshots from the front (or rear, as the case may be) that somebody had to think lateral to figure out that we could film gunfights, shots, and wounds without using a gun and without anything actually hitting the chest or body of the actor playing the "victim". This was done by simply pressure "bursting" a membrane at the tip of a tube carrying pressurized fluid resembling blood from under the clothes of the "victim" just at the time of the imagined shot (with a remote-controlled device switch)!

Once we become habitual of thinking in one way our mind relaxes and doesn't feel the need or urge to work hard thinking in a different way unless pressed or pushed naturally or by forced habit.

Almost obeying Newton's First Law of Motion, thought patterns also tend to keep following the groove they "fall into" unless acted upon by an "external force" of deliberate "lateral" thinking.

"Thinking out of the box" is, in fact nothing but an alternate expression for "Lateral Thinking"!

Although tradition and traditional ways are excellent auto modes of continuity of distilled wisdom and technologies, they also tend to hamper and suppress fresh thinking unless consciously opted for.

We have to learn and practice (lateral) thinking to escape the false solace and comfort of the status quo.

When everyone was comfortably and deeply sunk in the belief that trains were to be used only between far off places/cities and on iron rail tracks, somebody had to think laterally and propose the idea of running the trains on city roads as Trams alongside buses and cars thus making it possible for people to hop on and off near their places of work or residence without having to go to railway stations!

Lateral thinking can literally open floodgates of heavenly possibilities within the existing world!

5

Redefine, Reposition

"A properly defined problem is partially solved"
– John Dewey

By just changing the mission statement "to make agrichemicals" to "safeguard and boost the agri-produce" a manufacturer of horticulture pesticides, fungicides and fertilizers not only enhanced the umbrella of its products, it inspired both the farmer community as well as the company team to do more with a great zeal and higher purpose.

The sense of belongingness shot through the roof. The mission statement became a motivation statement as well! It brought down several imaginary walls of assumptions and restrictions in the minds of all stakeholders and made several inclusions in the company's portfolio possible.

Just a small stroke of a pen and a seemingly insignificant change of phrase altered the destiny of the organization and played a decisive role in helping it safeguard from future changes in the market ecosystem.

-

The soul of any product, service, concept endeavour or situation is in its definition.

How we define it, what definition we carry in our heads indicates what our approach towards it is! And approach determines the path and direction we take and hence the outcome!

In the words of John Adair,

"If the problem is incorrectly defined or analyzed, then it's almost impossible to recognize a solution."

Everything moves on the basis of some perception. A knife exists on the perception of either cutting vegetables or cables or for stabbing. The perspective is central to anything.

Change in perspective or perception of the purpose changes everything.

"The brief" is bed-rock of everything else.

It limits the scope, it liberates the scope.

Let's take an example...

What business are educational institutes in?

Running courses...?

Imparting knowledge and skills...?

Developing confidence, increasing employability...?

Preparing for a quality life?

Developing, nurturing human resource for higher achievements, potential realization and contribution...?

Or, transforming destinies of nations?

...the scope dramatically alters with just changing the business statement.

The paradox is that maximum business potential is as you alter the definition of its scope from conservative to liberal, from myopic to visionary...and as stated, you need vision, honesty, and sincerity for that! The poor cunning businessmen are getting peanuts courtesy their cunningness.

-

Imagine a beggar...as he defines his "work" in varying ways, each definition will have a dramatic effect on the "outcome"!

e.g.

"I am in the business of begging"

"I am in the business of making someone feel less guilty by donating a tiny part of his wealth or good fortune!"

"I am in the business of street entertaining", or

"I am in the business of being a stress sink for the society!"

--

What is your role? Change the definition and see the magic in self-motivation etc.

We generally limit the definition to "what we do". Instead, we should expand and raise the definition to "what we achieve or facilitate."

Like Theodore Levitt famously said, "People don't want to buy a quarter-inch drill; they want a quarter-inch hole."

People want a solution...so what ultimate solution you or your job offer not what actions you do?

--

Redefining the "thing" can dramatically change everything. It can expose wonderful possibilities.

It's like seeing the entire thing with different eyes, or from a different angle.

If the thing is not defined, define it for unleashing of magic.

Thoughts are not just a small train of words. They are very powerful, have big horse-power!

--

If it is already defined, try tinkering with the definition. Or better still, try redefining it....easier said than done, but worth it!

Repositioning tools can help redefine, though that is not the only way!

Just look at what mammoth fortune change was brought to the microwave sales in India by just changing the definition, the scope for the sales aim by repositioning the usage...

Around two decades ago, there were very few buyers of microwaves in India mainly because the product couldn't fit into any credible "usage" by the masses for some application they couldn't do without, mainly because of the fact that people just couldn't fathom to cook Indian dishes which involved curry, extensive sauté / tempering besides other complexities!

The researchers found a brilliant application by simply redefining the usage scope...

...reheating the cooked food!

They repositioned the microwave from cooking to reheating appliance. All restaurants and dhaabas were targeted successfully to reheat samosas, pakoras etc for every new walk-in customer (earlier they had to repeatedly heat up litres of cooking oil merely to heat up a few grams of the stuff, leading to big wastage of fuel!).

One of my favourite examples is from John Adair's classic "Effective Innovation" -

"At the end of the 18th century, Edward Jenner took the first step towards ending the scourge of smallpox when he turned from the question of why people caught the disease to why dairymaids didn't; the answer being that they were immunized by the exposure to the relatively harmless cow-pox."

This seemingly simple and small but brilliant change in the question threw open the door of a solution to one of the biggest pains of humanity!

Such innovation can throw open dramatic avenues for business opportunities and possibilities.

Edison's goal wasn't "make a working light bulb", but "make an electricity system cities can use to adopt my lights". Hawkins (the inventor of Palm Pilot) wrote a list of goals: fits in a shirt pocket, syncs seamlessly with PC, fast and easy to use, no more than $299. – Scott Berkun

6

Deep Thinking

In his classic "Effective Innovation" John Adair had said

"Creative thinking cannot be forced. If you are working on a problem and getting nowhere, it is often best to leave it for a while and let your subconscious mind take over. Your mind does not work by the clock although it likes deadlines. Sometimes the answer will come to you in the middle of the night."

Innovation is not always instant. Rather, it rarely is. It can be triggered instantly but it doesn't generally conclude immediately thereafter.

Innovative thinking and outcome are too brilliant to not to be beautiful. Beautiful things have to have a spark. And that spark comes, and it is not omnipresent. What may be always there is a flame of light, hope, and vision, but not that eureka flash!

In the advertising world, it is often fondly called "the big idea!"

An innovation is a fascinating, beautiful big idea!

Everybody believes in innovation until they see it. Then they think, 'Oh, no; that'll never work. It's too different.' - Nolan Bushnell

A hurriedly assembled or dug innovative idea can often be half-baked and not so beautiful, and in quite a likelihood ineffective.

Innovation requires patient waiting. It has to let the baby idea develop inside the womb of mind while floating upside down in the amniotic fluid of universal nourishment.

Deep thinking...

Many a time, you have to think a seed and throw it into the air and forget it. It is likely to come back to you at its will, time and place, but it will return. And when it does, it would have grown into a green-shoot, beckoning you to plant it!

> *"The test of a first-rate intelligence is the ability to hold two opposed ideas in the mind at the same time, and still retain the ability to function." – F. Scott Fitzgerald*

Thinking, like eating and praying, shouldn't be rushed. It should be natural and easy.

In John Adair's words

"Where there is a time delay, this means that the deeper parts of the brain have been summoned into action."

The world is flirting with the concept of data mining and machine learning while our mind had mastered heavenly capabilities in data mining millennia ago. All we have to do is provide it with input and leave it alone. It will dive deeper, wider and farther than we can even imagine and come out with astonishing results! The power of the human mind is yet to be fathomed more than the scratch on the surface.

While innovating, we have to often grapple with the facts, figures, possibilities, ideas, hunch feelings, observations, suspicions, urges, designs etc for sufficiently long periods of time before something radiant pops up! We have to let our mind do that for us. We have to master the process.

As goes the ancient Greek proverb

"While the fisher sleeps, the net takes the fish."

John Adair puts it this way –

"If you want to encourage new ideas don't evaluate too soon; give your seeds a chance to grow."

The techniques discussed in the earlier chapters are ways to artificially trigger the innovation process. Those are the maps to equip you when you go for treasure hunting. But when you are quite near the treasure but yet to "see" or "recognize it" you have to look with the mind! We have to give mind space and time to absorb and assimilate "everything".

Steven Johnson, in his beautiful treatise "Where good ideas come from" says

"...breakthrough ideas almost never come in a moment of great insight in a sudden stroke of inspiration. Most important ideas take a long time to evolve and they spend a long time dormant in the background....this is partially because good ideas normally come from the collision between smaller hunches so that they form something bigger than themselves.....eg it took Tim Berners-Lee around 10 years to get to the full vision of the world wide web....You have to figure out a way to create systems that allow those hunches to come together and turn into something bigger than the sum of the parts....space is required where ideas could mingle and swap and create new forms."

Creativity has unspecified incubation period which must be allowed. It is the time when the "raw materials" are allowed closed door magic to discover the magic and surprise us!

An innovator can be never sure as to which innovation process or idea needs time and space, when and how much.

All we have to do and can do is understand the process and the requirements which will hold our hands and guide us through the silent caves and jungles signalling to us when to keep quiet and listen.

> *"When what people do aligns with what they think and feel, then and only then, will you achieve the outcomes you're looking for." – Braden Kelley*

The more relaxed we are, the more submissive, obedient, understanding and silent we are, more our mind rewards us from the valleys of deep thinking.

Have a loose intention and set of 'hard rules', and within that, 'press play and see what happens'. Innovation is a creative process that needs some overall statement of an unmet need, but it doesn't need a detailed predetermined vision.

- Steven Johnson

7

Concept Extraction

One of the most challenging professions is being in a trauma center in a hospital. While any layman there is likely to go mad just on seeing the cases arriving, the team of doctors and support staff there are so cool (courtesy their training) that for them it is a routine which helps them use their head and steady hands in handling the cases.

Incidentally, one of the brilliant examples of "Concept Extraction" I came across is from the trauma centers. It is one of their routine procedures now but that was not always the case. Each time I recall it, I am simply astonished at the innovative acumen of the physician who must have done the procedure for the first time and that too under extreme pressure.

There are two paths of breathing for a human...one is through the nose (nasal passage) and another through the mouth (oral cavity). Both these passages meet in the back at pharynx leading to the lungs through the tube called trachea. Imagine an unfortunate condition when this airway gets blocked due to reasons such as extreme swelling, burns, foreign bodies stuck in there, injury etc. What do we do in such a situation? How will the patient breathe in such a situation?

Today we know the possible solution(s) but what all would the attending physicians have considered as options decades ago (if not centuries ago when this was first opted for)...they must have been under severe time pressure to make the patient breathe again. Maybe they even lost a few or several patients in the process and maybe they came out with a solution when they were forced to brainstorm while out of the operation theatre!

While a straight or normal solution to the problem would have looked seemingly impossible what could have made them see light at the end of the tunnel is – ponder over the root concept. A deep and serious thought would have revealed that the basic concept was "how to make the air reach the lungs so that the patient could breathe (again)." The basic aim was to have a path for the air. Now, if the normal paths were blocked and couldn't be opened (immediately), was there any other path available? If not, could another path be made available? That "Concept Extraction" would have lead to the astonishing innovation of making an opening through the neck into the airway bypassing the choked portion (known in medical terminology as tracheostomy or cricothyrotomy).

Today, whenever this procedure is opted for (there are a few other ways also innovated over the years now) the patient miraculously starts breathing instantaneously giving the much needed time to the medical team to work on the core condition.

--

Concept Extraction is an advanced tool for finding ways and spots of possible innovations especially when other tools are failing.

We have to go beyond the definition or questioning etc and straightaway delve deeper to the bone of the issue.

--

Let me give you another example to illustrate the "concept Extraction". This one, too, from the medical world.

Imagine a condition, like Acute Pancreatitis, when the pancreas of the patient is so severely infected that it is critical to give it rest for a week or more besides administering the antibiotics etc. But the main challenge in that condition is how to nourish the patient for that period when the only way to give rest to the pancreas is to stop eating orally.

Today we know the answer but I am amazed at the innovative breakthrough the initial physicians would have then found out for such situations, a brilliant example of "Concept Extraction". Well, they must have, in the course of thinking and grappling with the challenge, realized that ultimately all that was needed was nothing but making the nutrition reach the body parts/cells but without following the normal route of eating and digesting. What about pre-digested food/nutrition? But how to deliver it? How about directly injecting the same intravenous?

Brilliant! Isn't it?

Everything has an underlying concept which forms its bedrock. Whenever the normal innovation ways and means seem to be slipping, one of the fail-proof ways is to search and reach for the said underlying concept and apply the innovation tools directly there.

All you need is a deeper understanding of the matter with an untangled mind that misses or ignores no detail and discards no idea howsoever ridiculous it may seem.

After all, innovation breathes at the edges of logic!

8

Artificial Crisis

It is said that there is no real advancement without a crisis.

Or, that the best innovations happen under pressure against deadlines! Inventions and innovations during World War I and II are classic examples of this.

Use of Artificial Crisis is an acceleration tool for Innovation.

When all innovation efforts are leading to nowhere, this "third-degree" method is resorted to wherein we create a crisis artificially.

In this, what we fear, we manifest it ourselves.

We disturb the status quo...

For good or bad reasons and with good or bad outcomes (including collateral damages to the good and the bad) 2016 currency demonetization in India created an artificial crisis unprecedented in scale anywhere in the world and in the human history. This resulted in several forced, wanted / unwanted need-based innovations and adaptations in the eco-system including in the economic infrastructure, digital transactions etc. Overnight, businesses and people adapted to the new reality, which otherwise would have (in fact, had) taken over a decade to materialize! Survival mode was licked in and conceptual innovations were in top gear!

While evolution is an adapted and slowest form of innovation, a revolution is an artificially super-accelerated one!

You are thrown into the sea and you either learn to swim or find ways to survive, or perish.

This way of innovation is no doubt seemingly harsh and at times, immoral, but it either happens or is opted for as a last resort.

The balance of payment crisis in the early 1990s resulted in innovative policies of the then central government in India resulting in the start of an economic boom which lasted for over two decades!

In World War II, the crises post Pearl Harbour lead to the innovative though ultra-tragic tactical and strategic idea of atomic bombing of Japan which resulted in breaking of the deadlock and resulted in the premature ending of the world war which otherwise would have probably dragged on for several more years resulting in possibly larger tragedies and costs.

The famine crises lead to the use of innovative agri-techniques resulting in the green revolution in India in the second half of the 20th century.

This artificially created or naturally occurring mode of innovation may seem crazy, unwarranted, unfortunate, painful and unwise, it is a desperate attempt or opportunity to get a break-through, break the deadlock!

Though not to be blindly and recklessly used, its use becomes inevitable many a times.

In the words of iconic Guy Finley

"A crisis only becomes a **breaking point** when we fail to use it as a **turning point**."

Creating and using a breaking point to rally and harness reserve resources and energies in search of a turning point is a bold though crazy idea available.

***

9

Random

This is as amusing on the face of it as much as it is amazing when it shows its results!

When everything else is failing to lead to the innovation, this seemingly crazy tool is used.

Any random thing or concept or word etc is picked (especially something which has no connection to the subject at hand till miles) and "inspiration" taken from it.

It might sound rubbish or preposterous, but it works!

Let's take a (random) example.

Suppose we are to come up with innovative ideas for a missile. Also, assume that we are stuck up and try to use the 'random' technique to break the deadlock.

We look around and see a sparrow.

Now, how can we spark the thought process to come with an innovative idea for the missile using "sparrow" as the trigger?

Now, we focus all our attention on the sparrow and try to collect as many related facts as possible. Then we try to "see" the missile in the lights of those facts.

A sparrow is a small bird : Can our missiles be made smaller without compromising any of its abilities?

A sparrow prefers to live near human settlement and cannot be found in forests or deserts : Can missiles be deployed near the areas where they are likely to be needed? How can this fact about the sparrows be used?

Sparrows possess an extra bone in their tongue when compared to other seed-eating birds : Can we alter what missiles carry or are composed of?

In case of emergency, sparrows can fly at much faster than their normal speeds. Can we have speed altering missiles?

In addition to their flying prowess, sparrows can swim quite fast : Can we have missiles which can function as torpedoes equally efficiently?

Sparrows with their small weights and sizes can fit in any small openings : Can missiles be fitted and carried onto any other mode except the fighter jets or missile launchers?

Sparrows can build their nests mostly under roofs, or tree hollows : Where all can missiles be fitted strategically?

A small percentage of sparrow's eggs contain DNA of both parents : Can missiles have / carry any other type of "DNA" as well?

Sparrows live for about 4-5 years in the wild : Can missiles remain active or alive on solar or any other form of energy "in the wild" and strike / "cold start" from hibernation on its own when required?

-

If a random sparrow can inspire innovation for a missile, anything can inspire anything!

10

Accidental, Incidental

Though this pair-tool is not a deliberate innovation tool, I have included this here for the simple reason that these are too frequent and extremely useful for innovations.

In 1945, melting of chocolate in pocket led to the accidental invention of microwave oven by Percy Spencer when his alert and curious mind successfully linked it to the heat generated by the waves emitted by a magnetron nearby.

Tea bags were first used as packaging for loose tea samples.

While deliberate hard-work is the bulk of the innovation process, chance/accidental/incidental interventions cannot be ruled out. In fact, these are almost a norm. They look like luck but are quite dependable.

Wilson Greatbatch accidentally got the clue to reduce the size of a television-sized heart pacemaker to a tiny one small enough to be tucked in the patient's chest when he mistakenly picked up and used a 1-MegaOhm resistor instead of a mere 10,000 Ohm one.

Gorge Crum stumbled upon the idea to make potato chips when, in order to teach a customer who was constantly complaining about the sogginess in French fries, he cut the potatoes extra thin before frying them absolutely crisp. The hitherto annoyed customer was, however, delighted at the new recipe!

-

Whenever you start out to innovate, you can be sure to encounter some developments or ideas which are accidental or incidental...you never worked for them, you were not looking for them...you just stumbled upon them that they changed the direction and hence the fate of the entire journey.

Had pharmacist John Pemberton not prepared a drink with traces of cocaine to be used as medicine, we probably wouldn't be having Coke and Pepsi today (though, to be fair, these don't have the drug)!

-

In the words of Micah Soloman,

"If you can get your employees to be on the lookout for innovation potential in mistakes they've made and happy accidents the observe, it can pay off handily. If not, they'll continue to ignore the accidental leaps that occur—or, worse, bury these accidental improvements as being evidence of their, or their co-workers', errors."

However, accidents and incidents lead only the alert and sensitized mind to Innovations!

Otherwise, the apples must have fallen on a million heads before someone got the idea about gravity!

In Louis Pasteur's words

"In the field of observation, chance favours only the prepared mind."

11

Provocation

Big companies have trouble with innovation. Innovation is about bad ideas, or ideas that look like bad ideas. That's the fundamental thing. Ben Horowitz

This is a nasty tool...and, arguably most potent, dramatic and brilliant one.

This is what Dr. Edward de Bono calls "Po"...standing for the alphabets in the words...Hypothesis, Suppose, Possibility.

Under this, you come out with crazy (nearly bizarre) ideas which are visibly quite dramatic. So much so, that these are, on the face of it, quite disruptive and provocative for the interested parties or stakeholders!

This is akin to thinking the seemingly destructive or impossible or unacceptable way out.

Consider, for example, the shocking idea of injecting someone with the polio virus to prevent him from contracting polio! That sounds crazy and outlandish! But that was exactly the breakthrough idea behind the invention of vaccination!

Though it might seem reckless, this Innovative approach gives a formal green signal as well as the framework for some hitherto prohibited revolutionary thinking.

12

Assumptions

These are the landmines of the innovation field.

Every innovative idea once commissioned seems so obvious and logical in hindsight that at times, it is difficult to fathom as to why and how it was never thought out.

The culprit, as it comes out to be, is....assumptions!

Before somebody actually challenged this notion and thought about Trams, it was always assumed that the trains could move only in the wild expanse between distant places and not on the city roads amidst buses and other traffic!

Before touchscreen based phones appeared on the scene in 1992, everybody assumed that input could be made only through physical buttons!

Before spray painting was thought off, everyone assumed that painting could be done only by brushes!

Blame it on the way the human mind functions, but the fact of the matter is that our assumptions about almost everything in our lives are mind-boggling and beyond comprehension.

Before subway railway, it was assumed that a train was to move only on the ground. In fact, nobody thought of even any other possibility.

Not having any other consideration or idea is also indicative of the presence of assumption of the highest order!

For many years it was assumed (for whatever crazy reason) that refrigerators could only be white!

It is assumed that the pizzas have to be round only, pens have to be held between fingers, sleep has to be for eight hours, etc.

Spotting and challenging assumptions is one of the fastest and surest way to innovations!

It is assumed that TV's have to be either put on the table or mounted on a wall, or that they have to be something physical. It was long assumed that TVs have to be flat till concavely curved ones surfaced. It was assumed that surfing could be done on laptops and televisions were only for watching no-web programs!

Mere digging, identification, listing and understanding of these assumptions is, many a time, sufficient to throw open the innovations begging to be done!

These assumptions are so brilliantly camouflaged all around us that to identify them might be one of the most difficult challenges in the entire innovation gamut.

Till the Build-Operate-Transfer PPP (Public-Private-Participation) model was thought of by some innovative mind, it was assumed that roads (and other infrastructure projects) had to be built and paid for by the governments only!

Till digital watches surfaced, it was assumed that the watches had to have hands!

Assumption that physical wires were must to charge a device lead to the delay in wireless charging applications!

Assumption that the photographs had to be taken on the photographic films led to the sinking of the biggest of names in camera and camera film industry worldwide!

Till immortality is seriously challenged and there is really a credible way out, it is being assumed that the living beings have to die!

Before hovercrafts were conceived, it was assumed that the transport vehicles had to touch the ground or water while moving, or that the amphibious vehicles were not possible.

Before the electric vehicles, it was assumed that Internal Combustion Engines were a must for automobiles.

Not just the products, it is assumed that particular processes have to happen the way these have been happening. Those who challenge them find a better, efficient and breakthrough way!

Assumptions are blind spots in our thinking. They don't mean that there are no possibilities hiding there. In fact, there may be, in fact, have to be there infinite ones limited only by our imagination and belief!

Assumptions are notorious for hiding in many forms such as misconceptions, preconceptions, myths, apprehensions etc.

Often, they hide so conveniently behind comfort zones, temporary situations gradually cementing into pseudo facts over a period of time.

-

Industries rise and industries perish with the rise and fall of assumptions!

***

13

Rules

Just like a lot many potential innovations hide behind assumptions, there are quite a few which are hiding behind the "rules".

Rules of the game, the process, the procedures...

List down all the rules and challenge them fiercely.

The rule that the umpire's decision was final and there was no other way to it kept hiding the possibility of third umpire referrals for quite some time. But when it did, it revolutionized the game besides enhancing its credibility.

The rule that no help material is allowed inside examination halls in institutes is hiding so many reforms in the education system.

The rule that the turbo charger in a diesel engine has to be kept turned on at least a minute before moving the vehicle and after stopping it post a travel is a potential cover up of some fine innovation and breakthrough!

The rule that the majority has to agree for any law to be altered may be stopping some important legislation to be framed!

I am not saying that every rule must be broken and not respected. All I am saying is that rules are favourite hiding grounds and breeding corners of inefficiencies and assumptions preventing innovations.

Ruthlessly examine as to which of these rules can be broken! Which of these are unnecessary?

The result of the analysis is likely to be eye-opener. It is likely to throw the blind-spots and soft-spots out in the middle to be abolished or altered leading to breakthroughs!

***

14

NCs

When I was working in a manufacturing industry almost two decades ago, I got acquainted with the lovely concepts of NCs....short form for Non-Conformities.

I was in Quality Assurance department and whenever a product didn't match the designed or targeted specifications we used to term it as an NC – a non-conformity, the one which didn't conform to the laid and decided specifications.

An NC was a rejection, which needed to either be re-worked or discarded.

However, we gradually learned that NC was not as bad as it seemed. It was something not to be hated but understood and exploited!

Instead of just putting aside the NC product, we used to study it to find out the root cause, the source of the non-conformity. In the process, we got wonderful opportunity to plug the loophole in the manufacturing process which caused the quality lapse.

NC is a great opportunity to improve. It used to give us precious inputs about our systems and processes. It gave us hints as to where to look for improvements.

That is one of the reasons why I rate NC as one of the key tools for innovation.

Besides other tools, NCs / failures / rejections / set-backs etc can be great focus and starting points for innovation.

Instead of merely taking them as sore points which needed to be fixed, we should salivate to see opportunities for a thorough endeavour to innovate starting with that point.

Points of loss can be turned into points of gifted profits by giving us a clue about potential innovations.

15

Kaizen

You have to have a big vision and take very small steps to get there. You have to be humble as you execute but visionary and gigantic in terms of your aspiration. In the Internet industry, it's not about grand innovation, it's about a lot of little innovations: every day, every week, every month, making something a little bit better. - Jason Calacanis

Focusing on the failures / rejections / NCs to take corrective and preventive actions besides taking them as the high probability and high priority area of innovation may be an excellent approach, yet this more of a reactive approach.

Theoretically, it means – no failure – no improvement!

There has to be pro-active way to innovative at potential points.

In industry, we used to call it Non-NC quality improvements. We used to use various quality assurance tools such as 6 sigmas, TQM etc. But one of the most popular and effective tools was Kaizen, a Japanese concept which means "Small gradual improvements".

Years later, I thought of using Kaizen as a tool for spotting opportunities for innovation, in addition to the NC-based opportunities.

The concept is very simple. Just keep on systematically looking for small incremental improvements in every process, system etc and see all those as potential innovation spots using other tools discussed in this book.

This opened up a new avenue of potential innovation.

In those days in the industry, I had read somewhere about a unique Japanese practice of instituting 3 committees for improvements of various degrees.

While one committee used to come up with incremental improvement ideas, the second committee would come out with drastic / breakthrough ideas while the third one would come out with ideas which would render the entire present set-up, product-line, process and business etc of the company obsolete! Quite a harsh way!

But it is always better to simulate such scenarios ourselves rather than getting cannibalized by the competition or the market forces.

> *What is the cost of not innovating? What if a competitor launches a product or service before you do? What are the costs of being forced to respond, rather than forcing your competitors to respond to your great ideas? - Shelly Greenway*

In the words of Peter Drucker

"Every organization must prepare for the abandonment of everything it does."

Same approach of 3 committees can be used along with Kaizen based culture to innovate.

16

Quality

This innovation tool has also been imported from the manufacturing industry.

While quality, in the words of Phillip Crosby is nothing but conformance to requirements, it was also segregated as Quality of Design, Conformance, Performance and Experience.

If we were to dissect any process, concept, product, service, feature etc from the above 4 quality parameters, we would get another angle to slice it looking for potential innovative ideas.

By following the above or any other Quality model (there are quite a few) we can stumble upon several innovative ideas.

Eg. Quality of Design has three parameters – Features, Aesthetics and Serviceability. Focus on each one of these gives us fresh frames of reference to dive for Innovative ideas.

Dr.Philip Crosby quality model states that the performance standard of quality is "zero defects". This can form a good frame of reference to pursue potential innovation avenues. Another possible rallying point for innovation focus is Dr.Philip's concept of PONC (Price of Non-Conformity).

TQM, the umbrella approach to quality in any organization gives us another solid frame of reference that we can use to Innovate. Tools like 6 sigmas, Statistical Quality Control, 7 QC tools etc are also good triggers for innovating.

17

Problems, Feedback, Suggestions

An innovation will get traction only if it helps people get something that they're already doing in their lives done better. - Clayton M. Christensen

This tool is a version of NC tool discussed earlier, the only (though crucial) difference being that while NCs are detected in-house, problems are the ones shared or faced by the customers or users.

Problems, along with feedback and suggestions given by the users and stakeholders can give precious hints towards potential innovation ideas.

Around the mid 19th century, the typewriters had keys in only two rows and that too arranged in alphabetical order. this resulted in the clashing of the metal arm levers on which the characters were mounted, when pressed in quick succession especially when the successive character arms were placed adjacent to each other, e.g. 'st' (side by side) or 'th (first row 't' standing slanting ahead of 'h' in second row).

This problem was noted among others by Christopher Sholes who changed the design to a 3-row one with all vowels in the front row, thus solving it to some extent. The problem was almost completely eradicated by further changes by the Remington mechanics in 1873 leading to the QWERTY layout.

All problems are great hints to possible improvements leading to innovations.

OTP (One Time Password) was a brilliant innovation to solve the problem of unauthorized access of debit and credit cards.

Grades replaced Marks as an excellent way to curb stress caused in students because of excessive competition.

ATMs came into being to solve the problem of vending cash beyond and before the banking hours.

Pin-codes were designed to solve the problem of distinguishing between two or more places with same or similar names.

Innovation of car assembly lines happened to solve the problem of excessive obstructive movement of parts and workmen on the shop-floor. After this innovation, only the car sub-assemblies would move while workers waited at their respective progressive stations with parts to be fitted.

Third Umpire innovative idea came to solve the problem of controversial erroneous decisions which could mar the credibility of the game!

--

Problems are blessings in disguise, shrieking in silence about the innovations waiting to happen, the future waiting to happen!

While problems are generally to be spotted and acknowledged by the keen observers who are typically the insiders who have stake in the product, service or concept, these are also noticed and experienced by the customers and end users.

While this is a dangerous spot to be which may lead to collapse of the business riding on the product, service or concept, this can also be a potential money spinner and loyalty enhancer if the feedback (or complaints) from the customers and users is received with a thankful mind willing to work on it to innovate and improve.

Suggestions, feedback or complaints from customers can bring to forth hidden and invisible issues and improvement areas otherwise not in the awareness domain of the companies.

In the words of Mike Todasco, Director of Innovation at PayPal

"If you're being disrupted that means that someone else has come along and is solving a problem for our customer where we were not. For us that means we must never stop listening to your customers, we need to deeply understand what they need to improve their lives, and we need to strive to do that. Doing that well is the only thing that can keep us from being disrupted."

***

18

Observations

One of the very interesting Japanese concepts shared by Jeffrey Liker in his classic "The Toyota Way" was the "observation circle".

Under this, in a Japanese manufacturing unit, a circle would be drawn in one strategic location of a department and the new recruit would be asked to stay inside that for roughly the entire day.

While it might sound like a reprimand or taming tactic to the newcomer or any third party, the purpose of the unique exercise is to develop sharp skills of observation in the new recruit.

The person in the circle would initially find the "game" extremely debilitating and boring, but would eventually get so bored that he would inadvertently start observing the happenings around him.

Not doing anything and merely observing something non-stop for hours would result in the said person noticing certain things which a casual onlooker might skip altogether.

That's exactly the point. That is exactly what happens.

We are so engrossed in any environment, using a particular product or service, or running a particular process etc that the depth of our observation becomes shallower by the passing days.

This results in astonishing overlooks. All energies are gradually numbed to the level of mere maintenance of status quo.

This is where "the circle of observation" tool helps tremendously.

It helps you do deep observation resulting in the popping of precious innovation ideas.

> **"'Seeing around the corner' is a mark of great leaders."**
>
> – Jack Welch

19

Psycho/Functional, Gain/Pain, Constraint

This innovation tool is inspired and borrowed from the Sales function (though it can be applied to any function for innovation).

Sales and Purchase, of any product, service or concept is based on the promise of providing gain or reducing pain (whether tangible or intangible, functional or mere psychological) of the customer within his/her constraints.

In Innovation, we are always on the look-out for potential areas where the diamond ideas of innovation can be found and applied.

Using the above definition of the sales concept, if we were to identify and work on those gain/pain parameters, we could easily spot the starting points for potential innovation ideas.

Every product, service or concept has a purpose for some user, known or unknown which either enhances his gain or reduces pain while struggling to do so within the constraints. This challenge is a beautiful and powerful frame of reference to be used for innovations for that product, service or concept.

For example, if we are to find innovative ideas for a particular governance model, we can use the gain/pain/constraint model to find potential innovation leads.

Like feedback / suggestions, this tool too approaches innovation from the end users / customer's point of view but without the customer spelling it out proactively. It asks for deeper insights into what customer is thinking, believing, experiencing.

> **"An innovator is simply someone who can turn a new idea into a solution that can add value to a customer or to society," – Steve Basra, Toyota**

20

Competition

Reverse engineering is a very cheeky concept, seemingly unethical but unavoidable.

It is no-doubt, therefore, a crucial innovation tool.

Japan did it to Germany post second world war and China did it to both the East and the West since the turn of the last century.

They studied what the competition is doing and what the competition has already made, strip it down (product, service or concept), use it as a base for improvement, do the innovation magic and beat them back in their own game in their own backyards!

Not just the reverse engineering of the customer competitors' products and services but also its processes, best practices and everything it does.

It is an open treasure of fast-track innovation.

Not only from the established players, but also from the new "kids on the block"...

Mathias Herzog, Tom Puthiyamadam, and Nils Naujok, in a joint article in strategy-business.com had emphasized - "view each upstart competitor as a company you can learn from."

No company can afford to fall behind its competitor in any aspect.

Therefore, it is crucial that for innovation we turn outwards as well besides all the efforts invested inwards.

21

Future Already Happened

Innovation is the calling card of the future. - Anna Eshoo

Innovation happens at the edge of the time zone!

This is as much a reality as is a myth.

It is a reality because innovation is obviously synonymous with cutting edge features and technology.

A myth, because, at times, developments in the ecosystem have happened before these are adapted by surrounding products, services, systems etc.

Eg, several online smart-phone applications and services started much later than the actual technical feasibility of the same in the technical habitat.

Facebook, Whatsapp, UBER, Amazon etc could have technically happened much before that actually started. iPhone could have been designed and launched earlier than it actually did.

These launches appeared when someone conceived the idea to exploit the market with the already available technology.

So, spotting, identifying the future which has already happened is a great source of innovation. For this the fads have to be segregated from the sustainable trends so that we are in sync with the changes.

In the famous words of Bill Gates in his 1995 book (The Road Ahead)

"We always overestimate the change that will occur in the next two years, and [we] underestimate the change that will occur in the next 10. Don't let yourself be lulled into inaction."

22

Tactics

This is a more precise version of the "random" tool discussed earlier.

In my book "When Everything Else Fails,...Try Tactics" I have identified and shared **100 tactics** for adverse or hopeless situations.

Those same tactics or similar ones can be used to break the jinx and get inspiration for innovative ideas in any seemingly deadlock situations.

The purpose is not to find the logic but to stimulate imagination to see or use logic in an astonishingly refreshing way which is effective in achieving an objective which is otherwise stuck in a deadlock.

I am recommending tactics not only for the innovative design of products or services but also as a tool to achieve a breakthrough in finding innovative solutions to problems and situations etc. Innovation, as discussed in the myths is not limited to what the world believes to be.

Some of those tactics are randomly reproduced below for illustration and stimulation. Try adapting these to the challenge in hand:

Tactic 82: Shock overdo the unexpected

When the tidings against you come suddenly and at an unmanageable pace one of the best ways to arrest this terrible onslaught is to shock the enemy or the onlookers by overdoing and the unbelievable and unexpected act even if were to do damage to self (especially if the loss this way is much lesser than what otherwise threatens to be).

Whatever it results into, it does succeed in the prime intended objective.

Tactic 77 : Template multiply

What stands between a great idea and a revolution?
…….numbers!

All you need to create those numbers is a template and a multiplication factor. That multiplication factor can be media, word of mouth, assembly line, network, system etc...

You will be shocked to see what miracle you can do with the power to multiply in your hand!

Tactic 73 : Disturb the equations!

When the situation changes old solutions don't apply.
Even the smallest of factors can dramatically change the dynamics of the situation.

Tactic 71 : Exploit the SWOT

Only an amateur won't do this!

Everyone around you (and yourself) has a SWOT (Strength, Weakness, Opportunity, Threat) equation linked with him which keeps on changing.

Tactic 68 : Ride the opportunity

You want to escape from the situation. A horse is passing by. What will you do?

And the good thing is that a 'horse' is always passing by! Recognize it!

Tactic 66 : Ally

Befriend with those who share your beliefs and values and strategic interests.

This will be crucial to your survival amidst those who don't.

Tactic 58 : Drag in China!

Get someone strong on your side against the opponent.

Make the fight of your opponent difficult by several times.

This is exactly what Pakistan did when it gifted away some portion of the occupied Kashmir territory to China.

Tactic 53 : Neutralize Enemy's Strength

How ferocious really is the most ferocious dog with its mouth securely gripped close with steel net or, with his mouth free and open but with no teeth inside...?

A dog is nothing but its bark and bite! Minus both, he is less than a sheep.

Here the enemy may be anyone or any fact, etc.

Tactic 51 : Use

Anyone, anything, anywhere can be used as a tool in your scheme of things.

Tactic 45 : Noise

Noise is inversely proportional to clarity.
When you want to enhance clarity, reduce the noise.
And when you need to reduce the clarity and hamper communication, create or increase the noise.

Tactic 35 : Atmosphere, Ambience, Special Effects

How many times has it happened that the presentation, the packaging, the special effects, the atmosphere, the ambience did 'it' even before an analysis could be carried out on the actual thing?
Or, does this happen every time! I guess it does.
Although these special effects are no alternative to the real thing, yet they have a very strong influence on the fate of the real thing! They can bridge gaps! At times, they can even cover up the entire thing! At worst, they can smoothen out the edges.

Knowing that people decide emotionally and justify logically use the ambience and atmosphere to influence.

Tactic 17 : Get Big-time Endorsement!

Instant attention, instant credibility, instant recognition…..that is what you get when you get some big-time endorsement.

A word of caution: Choose the endorsement very carefully keeping in mind the end objectives.

Tactic 15 : Renovate, Rejuvenate!

Aging is certain, whether that of a person, or an idea or a concept or that of a product…

With aging, come dullness, repetition, monotony, and suffocation…

When the excitement of life or product or concept dies, it's the first sign of the approaching death or uncertain future. It definitely causes arrest in the growth.

Freshness revives that excitement. Renovation results in that rare combination of change as well as continuity.

It results in an all-new fresh inning, a fresh start. It even results in fresh ideas, a fresh stock-taking of SWOT, a fresh commitment, new goals, new meanings and motivations,……and it is the best time to shed some of the irrelevant, inefficient and non-value adding burdens.

Even the most successful of phenomenon require

contemporary redressing and facelift!

A word of caution: Do not disturb your core strengths, your core equations, your core values and value additions. Don't over-do it.

Tactic 11 : Do it in Steps

Eat elephant in slices - they say! That's right! Not only does it make the thing logistically feasible and psychologically easy but it also makes it a relatively invisible change. The resistance to such a change is highly reduced. The entire environment; including the human factor manages to cope.

Tactic 10 : Make stranger familiar & familiar stranger !

Very often the stalemate in the situation is because of the blind spots that get developed over
a period of time or general situation of lack of ideas. The perspective and the viewpoints get stuck up in such situations.

A whole new spectrum of ideas springs up when the hitherto familiar situation or person suddenly turns up as a stranger or vice versa...

Suddenly, the change triggers our mind into observation, analysis and action. We see many things which earlier simply went unnoticed because of the "used to" factor. Fresh ideas, fresh perspective, and fresh mind lead to a fresh approach to the entire thing. Besides, this generates a lot of excitement and "relate-to feel-good factor".

23

Convergence, Synergy

Innovation is taking two things that already exist and putting them together in a new way. - Tom Freston

There was a time when a television was a television only and didn't lock horns or compete with a tab or computer.

And then we saw TVs starting to do what was considered to be the forte of Tab and Smartphones. TVs became smart and offered to play a host of web-based apps such as Youtube, Amazon Prime, Netflix etc besides giving the power of web surfing in the hands of the viewer. On the other hand, tabs started streaming television channels too.

This is a beautiful example of convergence.

This beautiful clash is turning into a nasty war between the television and tab makers. Since both these sectors are overlapping increasingly, this was inevitable. Gradually, it is suspected and likely that they will become indistinguishable. One of these sectors will have to surrender. Many of the companies will either succumb or will merge. Consolidation and M&As are inevitable.

Wrist wearable touch screen smart mobiles are being called smart watches. This is a direct, fierce and unintended incidental assault on the watch segment by the mobile makers. Apple watch is leading the invasion while Xiaomi and others are also charging in. The high walls between segments are crumbling and the old citadels are about to fall.

Camera and Mobile segments are already at war with each other which the stand-alone camera companies don't seem to be winning, far from it!

Convergence and Synergy between two or more sectors, concepts etc is a fine innovation tool.

When absolute ideas are in short-supply, convergence and synergy are the easy preys innovators turn to.

24

Cross Application of Principles

This innovation tool is similar to the previous one except that this is the convergence or synergy between two principles instead of two products or services or concepts.

Eg How can the concept of democracy be applied to the internet?

How can traffic congestions be revolutionized using the internet protocols?

What can physicians learn from the experiences of the veterinary doctors (and vice versa)?

How about applying Newton's Laws of Motions to the pace of innovation in a company?

How can the Archimedes principle be applied to a sinking sick company?

What business lesson can we learn from the Season cycle?

What advice can we obtain from the principles of good photography?

What can the sports of Cricket, Football, Golf and Boxing teach us for application to businesses or any other field?

How can Darwin's theory of evolution guide us about the product life-cycle and that of an organization?

How can embryo/fetus development guide us about the development of an ecosystem?

-

This tool is one of the most exciting and potent advanced tools of innovation and helps break the stalemate deadlock conditions.

25

Analogy

"Earthworms that are cut in half can't actually regenerate from both pieces, but that we want it to be true shows that it's a great concept. Caterpillars metamorphosing into butterflies, owls seeing at night, and don't forget the humble mouse that became an essential adjunct to the graphic user interface (GUI)" – Micah Soloman

Many of the technical marvels and discoveries owe their existence to their inspiration from unsuspected sources.

Many helicopters were designed taking clues from insects. Robots are borrowing a lot from Humans. War games are getting clues from hunting tactics of nature's best group predators. Radar was inspired by the sound waves reflected from the bats...and so on.

Burdock burrs attaching to fur and socks with tiny hooks inspired the designing of Velcro.

The nose of the bullet train was redesigned like the beak of kingfisher to drastically reduce the sound it produces at super speeds and especially while emerging from tunnels.

Seawater desalination technology being developed by Aquaporin got its name and inspiration from the aquaporin molecules, a naturally occurring protein which facilitates water transport through cell walls that serves the important function of maintaining osmoregulation in living organisms. The amazing property of the aquaporin molecule to restrict the passage of contaminants, microbes, minerals, dissolved gases, salts etc while allowing water to pass through was exactly what was required towards designing a natural water filter for sea water.

Not only designs, analogy, mimicking, metaphors, simile etc are helping solve the problems faced by humanity.

Tardigrades, cousins of arthropods are inspiring the new technology to protect live vaccines while transportation to far off places.

Massive and complex power grids got their inspiration from the beehives to balance load during peak hours by remaining in sync based on technology mimicking behaviour of bees.

Discovery of Viagra was inspired by the ability of nitric oxide to stimulate blood vessels to dilate around the heart and other vital organs when released inside the body.

Any surprise that the aeroplanes from day one have been shaped like birds with wings! Isn't this more than a co-incidence!

Honeycombs are inspiring a gamut of areas including jewellery design and buildings.

Spider webs are inspiring breakthroughs in internet security and building designs besides other applications.

Advances in cryptography and block-chain technologies inspired the creation of Bitcoin!

The swimming center marvel that China developed for its 2008 Olympics (called Watercube) was inspired by the structure of soap bubbles giving it several advantages such as earthquake resistance, using sun's radiation to heat the pools, besides several design, maintenance and longevity benefits.

--

An analogy or inspiration (even if pretty odd and seemingly unmatching) can revolutionize the innovation process in any field.

26

Matrix

Demonstrated extensively in my book "The Matrix Window" a quadrant matrix can be a wonderful tool to logically slice any topic. It never fails to give us frames of references to approach any subject or topic which is giving no clue as to how and where to start.

Step by step, any situation or concept can be sliced into successive layers of 4 parts by making a 2 by 2 quadrant matrix between any two key parameters at a time. (3 by 3 and 3D matrices can also be used but those are a bit complicated. Also this 2 by 2 matrix breaks open any deadlock situation and its successive application generates enough momentum in the innovation investigation.)

This analysis can boost our understanding of anything throwing precious light on it making it easier to come up with innovative ideas.

--

A few examples are being shared below:

4 types of employee involvement

concerned about the company but not shared where the company is headed

= the company is undermining, ignoring and sidelining a big part of its internal strength. in tough times, this gap in the bond may prove decisive in the failure or success of an endeavour. the company is failing to leverage its strength and opportunities it offers. ulterior or hidden motives may be the only justification for not involving the employees but that is a clear hint that the growth of the company is capped. in that case, those asset employees are likely to one day realize the futility of seeing a long-term association with the company.

not concerned about the company and not shared where the company is headed

= this is a short-term uncertain association of relationship of convenience from both sides apparently and likely to fall apart in due course of time. the future of both is not encouraging.

concerned about the company and shared where the company is headed

= assets of the company valued well. employees, too, are on the right ship. good prospects of the company and everyone on board.

not concerned about the company but shared where the company is headed

= these are the cargo company ship must offload immediately to keep the vessel afloat. these are the black sheep which must be kissed goodbye asap lest the poison should spread.

4 types of market scenarios

closed economy, untapped demand

=suppressed economy, like India was prior to 1991

closed economy, tapped demand

=very powerful smart monopoly, mafia or dictatorship

open economy, untapped demand

=stampede, seller's paradise, quality not that high on agenda

open economy, tapped demands

=mature markets, quality and innovation back in driver's seat

27

Data Analytics & Pareto Magic

Pareto principle is a complementary tool to the Quadrant Matrix. This is a tremendous focusing tool. While there can be several innovative ideas competing for attention and implementation, Pareto analysis can help us prioritize the same.

This tool can be used along with other analytical tools such as ratios, fishbone diagrams, various statistical analysis besides input from machine learning algorithms.

If anything can be measured or recorded, it can be analyzed. Data contains messages and secrets of the phenomenon or subject in question.

It contains patterns, fault-lines, indications, suggestions, ideas, etc...begging to be seen, heard, decoded and acted upon.

> *"Data analysis is a process of inspecting, cleansing, transforming, and modelling data with the goal of discovering useful information, suggesting conclusions, and supporting decision-making." – Wikipedia*

In addition to the classic tools, the power of algorithms is being harnessed and unleashed by the field of Data Analytics.

This is being extensively and commercially being used in data analytics for predictive purposes, take better decisions, manage better, enhance efficiency etc.

All these classic and traditional tools are an excellent en-mass source of innovative ideas for the alert and willing mind which can see and sense.

28

The 10 Ideas Everyone Has!

This is one of the primitive, basic and old, though quite effective and popular method of generating innovative ideas.

This is inspired from the "Count to 10 and Win!" concept shared by Dr. Robert H. Schuller in his masterpiece

"Tough Times Never Last, but Tough People Do!"

though in a different context.

Before taking up the process of innovation for anything in a structured and professional manner, it is interesting to ask yourself (&/or everyone present) to quickly jot down 10 top innovation possibilities which instantly come to mind.

Before the minds are influenced and become too conscious, this is not a bad idea to take the cream of top points from their conscious and sub-conscious.

Many a time, you get 1 or 2 precious innovation hints from this exercise, often based on programmed gut!

In an innovation team of 7 members, if everyone contributes 10 raw, off–the-cuff, gut-feeling, top-of-the-mind ideas, you straight away start with 70 ideas! The final shortlisted or selected ideas may not be from this list of 70, but this initial pool will surely have some very crucial "informer", "fact stating", indicative, highly revealing and symptomatic hints. It will warm up the participants by good notches besides bringing to forth a good view of the existing realities and mindset!

29

What It Isn't?

This final tool for innovation is as interesting as the potency it hides in it!

Simply put, all it means is asking the seemingly innocent question

"What this thing isn't? And can't be? Is not expected or supposed to be? What can't be done?"

It is like looking at the thing (or concept) in reverse, opposite, behind, against, anti, inverse, virtual...

Eg. What is a spoon not supposed to do, not meant to do? What an aeroplane can't do? What can't be done? Etc.

While using all other tools discussed previously the aim was to have a frame of reference to help us think from various possible angles for innovative ideas.

This last tool is meant to make us think from the opposite side altogether...not about what all can be, but what can't be, isn't...to make us think everything else, anything and everything left out...

While all the previous tools were meant to help us see where and what all could and should have been seen, this tool is deliberately aimed at sweeping and gathering anything and everything which was on the side of our search vision.

It is a wild stroke (somewhat like the Random tool) with the aim that we get ideas also from the settings considered obviously impossible or improbable.

What all isn't there or can't be there throw light on a range of things which can be there. A good part of innovation is about looking at the impossibilities and improbabilities, and doing those things first!

To give a wild random example, it is obvious that car seat belt is not supposed to do anything other than holding us back against the seat. It is not supposed or meant to help us in driving, in navigation, in first aid, in massaging our tired bodies, in holding things for us, in talking to us and answering queries etc.

I see a smile on your face already on seeing this indicative list! That's the power of this tool!

***

Part C

Myths of Innovation

Myth 1

Innovation is Only for Product (& Service) Designing

This is the biggest of the myths about Innovation.

So much so that this myth single handedly has curbed the flight of the mega concept of Innovation in whole of human history.

Busting of this single myth can revolutionize Innovation and with it the present and future of humans and everything related to humans.

Innovation, till now, has been looked at as a mere commercial tool. A tool to win save or recapture the market.

No doubt, it is a very powerful marketing and commercial tool, but surely not merely that.

In fact, the myth only indicates that the innovative efforts have so far been implemented only for product or service designing. Or, most results have been recognized and acknowledged only for such applications.

In the words of Micah Soloman,

"...mental picture of Innovation is probably limited to the market-facing innovative leaps that grab all the headlines.

(Elon Musk's rocket flights, Alexa, etc.)...to expand this definition is essential..."

The wonderful liberating reality is that Innovation is universally applicable.

It is an approach which can help wherever help is required.

It is a thinking tool. A possibility searching tool. A creativity tool. A way to change.

Innovation is a problem solving tool.

You can use it on anything - any situation, problem, product, service, method, concept, challenge etc.

Whenever and wherever you are stuck, you can switch to creativity and its deliberately targetted effort we call Innovation.

Innovation is not just a designing or invention tool but a liberation door for and from anything.

"There's a way to do it better—find it."

Thomas Edison

Myth 2

Innovation Helps Only in Sales & Marketing

Companies are so much blinded by the immediate results and stock market sensitivities that they are unable to see or think anything beyond and in-between.

"The future us forever, but the present is this quarter" is a classic phrase as a testimony of this mindset.

Whenever people come to read or know that I am an Innovation Trainer, usually the first subtle reaction is, "O Nice! But we are looking for someone else. Our issues are not related to sales or marketing!"

By habit they link innovation with invention or break-through in product or service which, as per them, only the guys in Sales and Marketing need.

These guys may be from production and having a production issue, or from finance and facing a trouble in finance, or may be from HR and confronting a dilemma, or from Admin with a strange trivial challenge.

In the words of Mike Todasco, Director of Innovation at PayPal

"At PayPal we believe that everyone is an innovator! Innovation isn't just a function of R&D or engineering teams but is in all of us. It is critical whether you're in sales, legal or any other function in the company."

All I find them is either banging their head against the bad situation or looking from someone from their "field" only to see if help can be obtained.

Little do they realize that when facing the sticky and unusual problem in any field, you need to think out-of-the-box.

When you run out of the usual solutions to usual problems and get exhausted, you need to outsmart the problem.

You need to think creatively. You have to find an Innovative solution.

For that you have to look towards innovative techniques which you can apply in whatever field you are in.

Innovation is a field neutral field.

It is a way of thinking. It is a master-key for locks and doors of any building of any field.

From countering terrorism to rat-menace, from a stalemate in genetic research to finding solution to traffic situations in a city, from mental hacks to cyber threats.....Innovation can be applied anywhere.

Everyone needs it.

In fact, everyone deploys it, albeit consciously sub-consciously and in an unstructured amateurish random primitive way.

There are floodgates waiting to be opened by consciously welcoming Innovation in every arena.

Just like TQM, we need a revolution in TIM (Total Innovation Management) in every field/aspect on the planet.

"Innovation distinguishes between a leader and a follower." - Steve Jobs

Myth 3

Innovation – Only When Required

The key is to embrace disruption and change early. Don't react to it decades later. You can't fight innovation. - Ryan Kavanaugh

When, finally, people are convinced that Innovation can be a tremendous improvement and solution tool beyond product and service design, they fall into another trap namely "Ok! We will opt for Innovation when the need will arise. When there is a problem or a sticky issue!"

This is like missing the entire treasure and returning from the cave with a single diamond.

Or even worse, returning with the self-promise that "we will come and take it when we need it!"

Innovation is not just a problem solving tool but an Improvement and Breakthrough magic wand.

No doubt there is a limit to how much juice you can squeeze from an orange, but surely you wouldn't mind finding a way to produce more oranges in same space, finding higher yield seed, with bigger and juicier oranges, sweater...

Or, finding ways to preserve oranges for longer, ways to stack better, ways to reduce cost of production, packaging and distribution, ways to reach more markets, ways of overcoming cost fluctuations in markets, ways to find more applications of oranges other than juice, vertical integration of the business, switching to sister fruits of oranges etc.

Sky is the limit.

If there is no visible or on-hand problem, issue or bottleneck, find one! Don't wait for obsolescence to creep in.

"Adaptability and constant innovation is key to the survival of any company operating in a competitive market." - Shiv Nadar

Even if obsolescence is kind to you and spares, keep improving. There is no limit to new ideas and improvements with breakthroughs.

There was NO problem with the Blackberry, Nokia and other phones of the time when suddenly Steve Jobs came up with an idea to have a phone where you can click on the screen instead of a physical button.

A seemingly unforced un-required Innovation opened a multi trillion dollar business opportunity (while shutting out others).

This wouldn't simply have been possible had Steve Jobs waited for some "need for it to arise", a problem or customer grievance to pop up...

> **"You can't wait for inspiration,
> you have to go after it with a club."**
>
> **Jack London**

Myth 4

Innovation is done only by R & D Department

Too many organizations are so tied up in their organizational structures and processes that they are unable to recognize and harness the innate creativity that resides within their workforce. – Bill Fischer

This is as much an amusing myth as is a minor tragedy it represents with major consequences.

Once everyone is convinced (with God's grace and triumph of good sense over ego and closed minds) to let innovation in their lives, it is surprising to note that everyone expects the R&D department (or equivalent) of their organization to innovate for them.

Even those who are sufficiently excited about innovating are coerced by the tradition to let it be done by those who (supposedly) have been or are (presumably) deft at doing it or are "supposed" to do "it"!

Since innovation is no "it", "it" can't be effectively done even if it is done.

It is a deep rooted habit to assume that only R&D or some "specialist" department or group or designated persons can and will do the innovation for "others", that the responsibility of the "others" is merely to implement what "innovation" they are handed over.

Nothing is farther from the truth.

This thought simply indicates that the very essence and spirit of innovation is absent.

> *"The days when innovation was the preserve of research and development (R&D) units at the sidelines of the business have gone. Successful companies recognise that innovation is a mainstream process, which brings together frontline teams, customers and a range of different partners from beyond the organisation." – PWC (Unleashing the Power of Innovation)*

No doubt the innovation achieved by a core group can and should be multiplied throughout the organization wherever applicable, yet that core group has to be inclusive of all stakeholders.

Right solution can't come unless those involved in the solving process know the problem inside out.

This ensures the presence of all the right ingredients including the spirit, spark, understanding, ownership and passion.

Innovation is one thing which can't be delegated.

It must be attempted and done by "ourselves". Help of-course can and should be aplenty and on-call.

Only a lion and not an elephant or deer can innovate best for lions. Also, different lions are to participate for the innovation to include the varying realities within the lions.

The one who is at the heart of the matter has what it takes to conceive and bear the labour pains for the delivery of the innovative solution.

Nobody can or should do it for you.

Surrogate innovation is likely to fall flat on its face. Any such output is often without feet or vision. Not only is it difficult for "others" but also improbable that others will come out with solutions which will work.

So, innovation is a self-service for any person or group.

Every routine procedure you are doing should be open to the innovative eyes of the observer in the doer.

Nothing should skip observation. And no observation should skip questioning. Questions are the father of innovation.

"Finding opportunity is a matter of believing it's there."

Barbara Corcoran

Myth 5

Innovation is a One Time Affair

On the contrary, Innovation is an ever going process.

At the heart of the innovation culture is the belief that no situation or state of things is perfect, ideal or flawless.

Each point of reality is a seed with forests of possibilities inside it.

Reid Hoffman, LinkedIn founder and venture capitalist, phrased this brilliantly as "the importance of being in a state of "permanent beta." How to change mindset for this new age? "It's basically feeling that you always need to be learning. That you know things but don't know the whole game, and you are alert to how the game is changing."

There is always a scope for some improvement and a lot many breakthroughs in everything.

Just like a long walk comprises of numerous steps, though every innovation is one event in time it is not a one-time event.

In the words of Kevin O'Connor

"Innovation should be treated as a muscle that needs to be exercised; the more you use it, the stronger it gets."

The moment one innovation takes place, there are others waiting in the queue to happen, begging...

Also, one innovation doesn't mean there is no scope for improvement in the said parameters or situation thereafter.

On the contrary, an innovation is not the end of a road but a milestone...a knot in the climbing rope to stop falling below and to be used as a reference point to proceed higher.

Every innovation is a benchmark for the next one.

Innovation is a continuous process, a habit, a way of life.....full of thrill learning and rewards.

    "The best way to predict the future is to create it."

                                          Alan Kay

Myth 6

Innovation is Costly

It seems to be ingrained in businesses to ask, "What's the cost?" when considering a new innovation, but rarely do they ask, "What's the size of the opportunity we miss if we don't do this?"...Shelly Greenway

Yes, Innovation needs resources.

The biggest resource it needs is the minds to innovate and the process and skill of innovation.

Yes, It also needs monetary and other resources.

So, at the face of it, it involves a cost.

But, the cost of innovating is often peanuts before the treasures it can break open for you.

Just as Philip Crosby says that the Cost of Quality is the Price of Non-Conformity, the Cost of Innovation is the Price of not Innovating.

It may include the continuous bleeding of the organization with problems, inefficiencies, wastage of resources besides losing out on the future by continuously losing the opportunities and missing the must-turns on the way.

And we haven't yet considered the looming threat of extinction.

> *The very existence of potential disruptors in your industry — especially if they are funded by venture capital — is a sign that your business model is regarded as obsolete. It's up to you to figure out why, and how you can change it.* (strategy-business.com)

No company or nation or individual can bear the cost of not-Innovating.

In a recent article in Forbes by Bill Fischer, he had warned,

"The best response to the 'Our margins are too thin to innovate" excuse is – stay with this attitude and you have not seen anything in the way of thin margins, yet!"

All have been bearing it to their peril without realizing that the root cause was lack of Innovating.

Innovation is the hidden message in Charles Darwin's theory - "Struggle for survival and Survival of the fittest."

In effect, not only is Innovation free, it is effectively ridiculously free!

Innovation is the prime source of sustained profits.

As a consolation note, however, it is pertinent to be added that at times instead of arranging or allocating fresh or more resources for innovation, efficiency enhancement can free up part of existing resources to be used for innovation. Therefore, innovation should first be employed internally to free resources to fund itself.

"Genius means little more than the faculty of perceiving in an unhabitual way."

William James

Myth 7

Innovation is Disruptive

Yes it is, but not in the negative sense it is instinctively implied and taken.

Innovation, at best, is relief imparting and at worst, very disruptive.

But for once, self-induced disruption is better than competition- or environment-induced disruption.

The former can be life-giving labour pains while the latter can be our assassination.

In the words of Micah Soloman,

"Before your product or product line becomes the victim of the next wave of Uberization or Amazonization in your industry, encourage employees to look at what's missing in your company offerings themselves, even if means questioning what they may think are sacred, untouchable cows."

The beauty of innovation is the tremendous value unlocking by a simple (yet, at times, astonishing) brilliant solution which packs the punch of a cluster of several otherwise complicated traditional answers.

In other words, innovative solutions are respected and awed because of their low cost mind-blowing value delivery.

And this quality is precious because they are somewhat disruptive.

Or, said in reverse, since they are disruptive, they are notoriously effective and efficient.

You do have the choice, though, to opt for non-disruptive non-innovative incremental improvement route. But then, where improvements start failing and slipping, innovation giants start to wake up.

Myth 8

To Innovate, You Have to Be a Genius
& Very Creative
It Is Not For Everyone

The fact that talent is quite low down the list would further underline the move from innovation being 'alchemy' by the few to 'cookery' by the many. – PWC (Unleashing the Power of Innovation)

What a myth!

How mischievous!

Did they spread this rumour deliberately to stop the masses from innovating and thus nipping any competition in the bud?

Or did the masses themselves surrender with this beautiful innocent excuse?

You have to, however, be a genius by not believing this myth.

This myth needs to be blown away with dynamite!

The main problem with the masses is that they don't know how to think.

The technology of thinking is so simple that if that were to get revealed openly, everyone and anyone could be a "genius" and "creative"

All you need to know is "how to"!

On being asked whether Innovation could be learned, Mike Todasco, Director of Innovation at PayPal said

"The top innovators we award every year are not like the eccentric inventors or introverted scientists that you might see in movies. They are curious people who are excited about creating great things for our customers and really work on their innovation skills. Grit and learning from mistakes are the two things that I see the most in those people."

Creativity is as basic a circuit in the brain as it can get.

Only, it has to be dusted and switched on.

As beautifully put by Skot Carruth –

"Why would a smaller, less experienced team be better positioned to solve a problem that the current one couldn't? The answer is Shoshin, a Zen Buddhist concept that translates to "beginner's mind." Shoshin describes an attitude of openness and curiosity, a generalist mindset that allows someone to approach problems without self-imposed limits."

Anyone and everyone can be creative and innovative. Just need training.

"There is an innovator inside of all of us."

Rowan Gibson

Myth 9

Only Customers are Interested in Innovation

Innovation comes from the producer - not from the customer. W. Edwards Deming

As discussed, Innovation is not just a product and service design tool but a survival, problem solving, continuous improvement and breakthrough tool.

It is applicable to anything and everything under the sun, including every aspect of our daily lives, situations, concepts, tangible, intangible.

So, the age old myth that only "customers" are interested in innovation is much outdated.

It acknowledges innovation's role only in "marketing" objectives. It fails to see the other 99% of applications of Innovation.

There are numerous domains where innovation turned hells into heavens (in mind or in the world) but couldn't be recorded in monetary terms in some balance-sheets.

Who is a "customer" in broadest of sense? Everyone is...Who is not?

So, everyone is interested.

In fact, everyone will and should be interested in the tangible and intangible rewards and benefits Innovation has to offer to them.

The only question is how many out of those "everyone" are aware or have been told about their "right" to the fruits of Innovation.

The simple process of focusing on things that are normally taken for granted is a powerful source of creativity.

Edward de Bono

Myth 10

Innovation is Futuristic, Luxury for Present

Can't stop but laughing at such mindsets (albeit with a tear each in both eyes)...

Just because Innovation has been always treated as something futuristic and luxury, it has been ignored and pushed aside only to cover the ignorance of "how to innovate".

This single excuse and alibi have robbed the world of mind-numbing improvements and breakthroughs which could have, to a good extent, already helped alleviate pain and poverty from the planet besides having already opened a million doors to heaven on earth.

The cascading repercussions of this myth are mind-boggling and unacknowledged.

Innovation is a problem solving, improvement, breakthrough tool.....a possibility harbinger.

In the words of Amy Sorrells –

"The first step in solving a problem is recognizing there is one and the signs are everywhere."

Innovation is not a luxury. It is a sheer basic necessity, in fact, an absolute basic one...

"The secret of change is to focus all of your energy, not on fighting the old, but on building the new."

Socrates

Myth 11

Innovation is Difficult, Takes Time
(Like Invention)

Very innocent myth indeed...

Within the question lies the answer, within the wording of the myth lies its root cause.

Since childhood we read that so and so thing was invented by so and so scientist for which he got a Nobel.....and so on

The very fact that we equate an innovation with an invention, we equate the innovator with an inventor.

Therein lies the whole trouble.

While innovation and invention can surely be taken as synonyms, commercially and figuratively they shouldn't be.

An "invention" is a celebrated branded huge creation, while any new idea, thing or way can be an "innovation" even if tiny casual and seemingly very simple.

To repeat what I shared in Part A,

"While invention is the creation of something for the very first time, innovation is a sort of recreation of it albeit with a significant difference. It is like creating the same again (with clues from the existing one permitted liberally) in a different way (partially or wholly). While the thing essentially remains the same, it carries a sense of newness about it in whatever way!"

While invention is perceived as a very scientific complex and difficult phenomenon, an innovation is almost a commonsense which was previously either not visible or not seen or simply overlooked...a potentially everyday non-event of very high importance.

While it is a perception that for an "invention" a scientist is required (this is, again a myth though), it is increasingly dawning that for an "innovation" any human who can think consciously can be the one.

An invention which is done by a common person without much ado is an innovation.

While the above discussion was to emphasize that Innovation inherently is not repulsively "difficult", it should also not be construed that it is so easy that it doesn't demand much of application and we are doing some sort of favour to it for letting it happen.

Two essential things decide whether the Innovation juices in your mouth greed saliva or hunger sings.

1. The Mental Perspective :

    As goes the famous expression that the mouse always wins the cat-mouse chase because while the cat is running for her food, the mouse is running for his life!

Are you are the proverbial mouse here who is running for Innovation as a life-line or the Cat who is after the Innovation for some kind of recreation or incentive!

Perspective decides the outcome as well as its speed and shine!

2. The Passion :

Absence or presence of "Passion" is clearest of signals of fate of the impending future of any venture, including Innovation.

-

The perception that Innovation takes time is also not right. While at times it may take more than anticipated time, most of the time it is much swifter than the "Invention story".

The tools and techniques discussed in this book are for this very purpose only, to use templates to reveal possibilities, help break the deadlocks, trigger ideas and pop up innovation green-shoots which may be lopped up by alert and eager minds and enlarged into full fledged Innovations!

Innovations per se take only a fraction of time an Invention is perceived to consume!

Moreover, it has to be fast failing which it loses its value.

In Gerry Collins' words, "Time is the key metric for Innovation – how fast you can move?"

In Kate O'Keeffe's words, "By the time you have connected with end users, industry players, investors, subject matter experts, marketing, engineering...it's almost time to start the whole damn thing again because no doubt the market will have moved."

The turnover time from the realization of need for Innovation to the actual implementation of the innovated Innovation followed by results has to be very less. Otherwise the entire purpose is defeated!

"There is no doubt that creativity is the most important human resource of all. Without creativity, there would be no progress, and we would be forever repeating the same patterns."

Edward de Bono

Myth 12

Innovation Is All About A Dashing, Flashy Big Idea

Most people think innovation is about ideas. It's not. It's about identifying and solving meaningful problems. That's why simply coming up with a new idea won't get you very far. – Greg Satell

Well, well, well....

While many ideas are very dashing and breath-taking, not all have much utility under the constraints!

Therefore, while this myth is not always a myth, it should be treated like one because while not all flashy ideas can be called useful innovations, there are many innovations (and potentially budding innovative ideas) which are not so flashy, are pretty simple and ordinary at the face and are therefore at the risk of being taken lightly or out-rightly ignored.

Innovation, like a genius, is 1% inspiration and 99% perspiration. Had the world waited like Archimedes in the bath tub or Newton below the apple tree, the scenario would have been unbelievably different today! (That both Archimedes as well as Newton were able to spot what they did, in fact was also because of painstaking preparation of their minds instead of having been caused by some unexpected flash!)

While majority innovations are brilliant, dashing and often flashy once they sink in our understanding, not all flashy ideas are innovations. While they may be unorthodox, surprising and even thrilling, they may not be feasible (I am not saying practical). In other words, they may either not be optimum solutions or not solutions at all. It is the final utility of an innovation that counts and not its beauty or brains.

In the words of Soctt Berkun,

"When inventors are asked how they came up with their idea they say "it just came to me", and have some interesting story about how the idea formed in their head. They don't say "after 1000 hours of research the idea became obvious", because it's boring. Stories about epiphany's are interesting and exciting and give us hope that we will have on."

We are at the serious risk of missing some of the most effective and brilliant ideas if we start judging or rating an idea by the template of flashiness.

To carry brilliance of utility, mindful of the hidden demands of the situation, of the innovation gets ensured only when the innovator first thoroughly studies the situation / problem / concept in hand.

It is the inherent power of an idea and not the superficial or functional flash in its beacon that decides or should decide the value of the idea.

Spare innovation the glamour glare (atleast till the pre-market stage) and you are doing a big big favour to everyone including yourself.

If your main weakness is idea selection, or idea execution, then generating more ideas won't help. In fact, generating more ideas can actually make you less innovative, because the weaknesses in other parts of the process will sink the new efforts, which in turn increases the frustration of your people – demotivating them. – Tim Kastelle (hbr.org)

Myth 13

Innovation Is Too Fancy A Thing To Be Of Interest To Management

(who have more important things to do)

"The problem is that while the eyes of the CEO are fixed on innovation, the body of the organisation may not be following. The 'antibodies' that inhibit innovation include a culture that sees it as separate from the mainstream operations of the business." – Unleashing the Power of Innovation (PWC)

I couldn't stop smiling for some minutes when this myth first danced before my eyes.

Managing is a serious business, without doubt.

There are hell lot of pressures.

Hell lot of money is at stakes and so is the future of all stake holders more so of the promoters whom the top management is answerable to (if not already a part of it).

The day to day grinding challenges and responsibilities of the managements are too important and priority "than being a spendthrift of precious time on something fancy and flashy like innovation where nothing concrete is assured or guaranteed!"

This is a disastrous and tragic thought.

Innovation is a company-wide tool....a basic and first tool of business and everything else...because it is the active form of thinking. What is management if not thinking (and action based on that thinking)?

We have been erroneously equating innovation as something glamorous, flashy and fancy and hence this mindset.

In whatsoever the situation, innovation should be the first thing management should lay its hands on. Like a police commando whose natural instinct is to reach for his pistol in the holster whenever there is even a faint hint of danger or need.

Though Innovation's effects are fancy, Innovation per se is not fancy or flashy.

Innovation is very simple basic and serious work.

In fact, without innovation a management may find a solution which is costlier, less effective and even an out-right blunder.

In that situation, the seriousness demanded by the Innovation (but ignored by the management) will haunt the (so-called) "serious" management for a good time!

Nothing is more important for the management to do than to manage the health and longevity of their company. And what better management tool for that than Innovation (small or big, anywhere, everywhere!)

Having said that, it must also be ensured that the pendulum doesn't swing to the other extreme (equally dangerous) end, namely, the management taking all the glorious burden of "Innovation" on itself...

As put straight by Micah Soloman,

"Although innovative leaps do sometimes come directly from the CEO, this is less common than business profiles and biographies make it seem. More commonly, successful leaders drive innovation via their people: by inspiring and sustaining their employees..."

"Innovation is the central issue in economic prosperity."

Michael Porter

Myth 14

Innovation has Cascading Effects / Implications

Well, this is not a myth per se.

This is true.

I added this to the list of myths because I wanted you to be sensitized about this aspect before you misunderstand it or are scared by it leading to an inevitable, dangerous and costly myth.

Innovation is brilliant because it often finds simple, economic, often surprising and amusing, at times stunning, effective solution to a very stubborn, sticky or seemingly dead-end impossible problem or situation.

Because of its inherent lateral guerrilla-warfare nature, innovation often results into changes which are for good, very good.

Mathias Herzog, Tom Puthiyamadam, and Nils Naujok (in their November '17 brilliant article in strategy-business.com - "10 Principles for Winning the Game of Digital Disruption") gave principle no. 1 as "Recognizing the change: Embrace the new logic."

They go on to aptly warn -

"If your company is already struggling, then digital disruption will accentuate your problems. You may not have needed a plan for the new digital age yet, if only because it didn't seem relevant to your industry. But you will need it now. Otherwise, no matter how well you run your business, it will not produce results at a scale that will allow you to compete."

Change may demand stepping out of the comfort zone. In fact, 'being comfortable with your un-comfort and uncomfortable with your comfort' is the whole essence of Innovation.

It may result in some implications which may be taken as side-effects for the established regime of things.

It is human nature to resist collective good if it is bad individually.

If the change is disturbing, painful or uncomfortable (which it often is), it is often not swallowed willingly with a smile.

It is resisted till the pain of not changing is decisively more than the pain of changing. But by that time, often, nothing substantial is left to safeguard or achieve and collateral incomprehensible loss has been done, beyond knowledge.

Innovation may demand readiness to the changes / effects / implications. At times, astonishing innovations open flood gates of bright future thus resulting in a short term and temporary dislocation/relocation/re-arrangements etc. But it is a fraction of the price for what we get in return, where we reach or the road we land on...

However as a note of caution, not all innovations are unpleasant. Many are very sweet and very comfort imparting.

From that point of view, this may be taken as a half myth.

All i wanted was to sensitize you on the point.

"Do not fear to be eccentric in opinion, for every opinion now accepted was once eccentric."

Bertrand Russell

Myth 15

Innovation Creates New Problems

Well, it may. It is likely to.

There is no guarantee that any innovation will be the last one in the series.

Innovation is a solution or progressive idea in the backdrop of a recognized problem or opportunity or a situation. It is naive to assume that no situation or problem or opportunity will be there till eternity.

In fact, any solution is the starting point for the next innovation, improvement, breakthrough...

Future, once arrived at, is the present which forms the base for the future thereafter.

An innovation which results in problems is a half baked solution, if not a bad or ill-selected one. It may also be that it was wrongly implemented. There can be many reasons.

But a good innovation solves much more than the new challenges it creates, instantly or later on.

It may also happen that an innovation solves a huge problem but in the process throws up a challenge of solving some new (though much smaller) problem to benefit from the main large solution offered.

Who wouldn't mind a pound of solution by facing a conditional consequential intermediate penny problem?

The "problems" or difficulties or disruptions a good innovation creates are a fraction of the benefits it brings, the solutions it offers.

In the words of David Kadavy –

"If we never get started, we never get good, and you can't get good without first being bad."

Innovations need adjustments resulting from the implementation of the idea. While several are small and are met willingly with a smile, some may demand further intervention to set right.

We can't say no to a much bigger good which has a new tiny bad as a neighbour on the way. Who knows that bad neighbour be a god of some new innovative breakthrough?

"The stone age didn't end because they ran out of stones."

Anonymous

Myth 16

Innovation Happens Best Under Severe Constraints

Yes, and No.

While it is fact that we are forced to act when our back is to the wall or when we can no longer procrastinate or close our eyes to a situation, and while it is also true that the bad or challenging times often extract the best out of us, it is also equally true that some of the great Innovations happened out of conscious efforts or without any threat or pressure, under utterly comfortable and inspiring scenarios.

This is, therefore, a myth which we must caution ourselves against.

Otherwise, there is a strong likelihood that we may keep waiting for the pressing need to arise, a crisis to unfold before we take note and start to go hunting for an innovation which may be too costly and late.

It is very likely that we may miss out on all the progress that Innovation in peace time would have offered.

While Innovations around the problems are merely a reactive subset, proactive Innovations around the possibilities can be innumerable much outnumbering the former.

However, Clayton Christensen's following words too need to be kept -

"Large organizations don't go far enough in the pursuit of non-core innovation, in part, because they always must keep an eye on the core business."

Having said that, in Innovation, "don't fix it till it is broke" maxim doesn't apply.

Rather, the exact opposite is the Anthem.

> **"There's a way to do it better—find it."**
> **Thomas Edison**

Myth 17

Innovation Needs to be Managed

Nothing can be more dangerous and toxic for Innovation than "managing it".

Innovation hates management. Innovation doesn't need to be managed, Innovation process needs to be. And that too, with a very light hand, and from a distance…

In the wise words of Jeffrey Phillips

"Managing Innovation is at best limiting Innovation, at worst, blocking it. Rather, what we need are enablers, people and systems that understand that innovation is a bit random, messy, uncertain, and cannot be managed the way proven, described and repeatable processes can."

Innovation is a fragile and uncontrollable process. And that is what ensures its purity and brilliance. Without that, it remains an old helpless lackluster status quo.

Innovation needs pampering. And in return, it rewards you many times by giving birth to the future which was about to be lost in the complacency and arrogance of the present.

Myth 18

Least Risky Innovation is the Best

There is no innovation and creativity without failure. Period.
- Brene Brown

That is not an "Innovation", that is an alibi, an eyewash, a consolation.

Innovation is a flight of imagination. It may be guided or triggered and artificially boosted. Still, it is a tremendous flight of imagination.

It is an endeavour to think out of the box (even if guided by the box), break the walls / cages / restrictions of myths and assumptions...It is an attempt to think and consider what was never in the domain.

Newer ideas, new approach...

And when it is this much of an adventure and wandering for some brilliant discovery, the outcome is never pre-determined and neither are the contours of the journey.

Therefore, to say and expect that there should be no or minimal risk is to turn off the throttle, choke the breath of creativity, play safe and yet expect a breakthrough!

The most resilient companies foster a pervasive culture of innovation at all levels of the organization - one that values

risk-taking, embraces experimentation and considers failure an inevitable part of thinking boldly. - Lynne Doughtie

Innovation inherently is a "risky" venture. But we have to acknowledge that "nothing ventured, nothing gained!"

What we fathom as "risky" is nothing but fear of failure hiding as lack of confidence to venture out of the box for fear of the unknown there.

It is like starting with a pre-conceived notion, a bias, a prejudice. It is like starting with a rough idea of what you are looking for.

"Why" you are looking for the answer to a particular problem or objective is acceptable, not the "What", because if you start with that mindset, you miss all that is not in the focus of your pre-decided mind!

'Failing fast' is an important component of cultivating the agility needed for a development project along with continually rebalancing your innovation portfolio. - Lynne Doughtie

Myth 19

Innovation Guarantees Breakthrough in Business Success

In most parts of the world, starting a company that goes bust is dubbed a 'failure.' In Silicon Valley, we call this 'gaining experience.' We are willing to take the risks that are inherent for innovation. - Sebastian Thrun

Surprising but not final and absolute truth!

While there can be little breakthrough in business success in contemporary times without Innovation, and definitely highly unlikely in upcoming times, the reverse is also an eye-opening truth.

Put in simple terms, what it means is that Innovation itself might turn out to be a damp squib if customers don't appreciate it or want it. Innovation has to be in sync with the times and with some customer gain or pain under his constraints!

There have been numerous examples where the organisations, individuals etc came out with a very promising innovation but either didn't or couldn't use it, didn't believe in it or it simply didn't work out all the way for inherent reasons.

Scott Berkun gives a brilliant example in his "the Myths of Innovation" –

The Japanese invented firearms years before Europeans. But their culture saw the sword as a symbol of their values: craftsmanship, honour, and respect. Despite the advantages of using firearms, the innovation was ignored and seen as a disgraceful way to kill. Innovations do change societies, but they must first gain acceptance by aligning with existing values.

Ulterior motives and reasons are often known to hinder innovation process. Culture, comfort, beliefs, constraints, political compulsions, protectionism, policy frameworks, stakeholder pressures, etc impede the natural course of innovation.

Many countries are still sticking to the imperial system of units of measurement while sweeping aside all the logics and benefits of the metric system for this very reason.

Kodak already had the digital photography technology in its hands but couldn't muster enough courage to go against its bread and butter analogue technology which was still going good. By the time they shrugged their fears off, it was too late!

As shared by George Baroudi, VP for Information Technology at Long Island University, "Google has been more Innovative than many companies but most of these innovations end up staying as just a good idea. Let's face it - every developer would love to have their own incubator. Amazon is more focused on consumer needs, and therefore developers do not have an opportunity to work on ideas that may never manifest themselves. The difference between Google and Amazon - Amazon is a pragmatic Innovator and Google is a blue ocean innovator."

It has also been seen that it is also in the interest of a part of the establishment for the status quo to continue. If that part commands influence more that the faction pro-reforms and innovation, any change gets vetoed, openly or diplomatically and subtly. Eg there are several who are not interested in gun control measures in some countries. A good part of the automobile industry barring the likes of Tesla etc have been "blocking" or "going slow" on electric (or alternate fuel) vehicles for the fear of the entire favourable eco-system based on Internal Combustion Engine collapsing, making way for the much more efficient and capable electric one.

Examples are littered all over. Breakthroughs caused by Innovation are at the mercy of soft and hard powers of the environment, perspectives, decision makers and eco-system besides several hidden and innocent reasons and causes many of which never come to light!

Myth 20

Innovation is an Event

This is a classic misconception.

Innovation is not an event even though it appears like one.

It appears so because of the glamour linked with the Eureka moment that a typical dramatic innovative idea is supposed to culminate in causing a breathtaking disruption. The fact that every innovation doesn't make such a sound is another issue.

In reality, Innovation is not an event, but a process.

The discovery of an Innovative idea is not an Innovation in itself; far from it.

By then, it is just a skeleton of it, the core. It needs to be developed besides the need to make the necessary preparations to implement it. Many innovative ideas need elaborate work to be done before they can be put into action. Not to forget the intermediate testing, prototyping or trial stage.

Several innovative ideas are also known to go bust before one in the series finally comes good.

In the words of Greg Satell,

"The gap between discovery and commercialization is so notorious and fraught with danger that it's been un-affectionately called the Valley of Death."

He goes on to say that

"You can't really commercialize a discovery, you can only commercialize a product and those are two very different things. The truth is that Innovation is never a single event, but a process of discovery, engineering and transformation."

Having said that, it is not improper to give special status and treatment to the Innovative idea because it generates the motivation and the thrill required to undertake and complete a challenging transformation against hard odds.

Myth 21

You have to Innovate Yourself

Well, well, well...

Innovation is Innovation...an innovative idea, concept, design, business model etc. which has a wow factor of dramatically solving our problems by enhancing our gains or diminishing our pains. Whether it was dug out by us or simply fell into our laps is immaterial.

Provided, of course, we are passionate about it and have the right mental perspective and attitude. We are hungry for it and respect it.

Take, for example, copying of the innovative business model of UBER by OLA. Nothing wrong with it... Similarly, FLIPKART did an Amazon in India, and OYO an AIRBnB!

Ignor Stravinsky had said - "Lesser artists borrow, great artists steal!"

I am not promoting or endorsing violation of Intellectual Property rights...far from it!

But if an idea is not copyrighted, geo-spreading it or using it is not a crime. Not doing it is...

Just because a company has made an automobile, television or PC for the first time doesn't mean others can't as well.

Competition's response is its version of the Incumbent's move. Reverse Engineering has been a reality for decades if not more! Whole of China's mega success story exemplifies that astonishingly!

"Sustaining Innovation" of Innovation too is a valid, necessary and inevitable phenomenon.

Digging a well when there is a gushing stream flowing by the side (unless prohibited by law or other genuine reason) is wasting the capital and stretching the luck.

It is not a compulsion that you have to innovate yourself. You can outsource, improvise, alter etc.

Just because you can innovate yourself doesn't mean you should be ignoring, disrespecting and trampling under your feet precious ideas which you can employ with an edge.

Having said that, two things must be kept in mind...

First, even when you are merely using an already-done Innovation which is under-exploited you must give it the treatment worthy of a fresh Innovation.

Second, just because something is cheap doesn't mean it will be economical. Innovation idea should be adopted only if it strategically suites you and doesn't compromise your competitiveness in any way. Once adopted, the Innovation becomes your baby and treated accordingly!

Myth 22

You "do" Disruptive Innovation

You don't "do" Disruptive Innovation.

It happens!

An innovation is an innovation. Before it actually happens you are never sure which type of innovation it will be. (see chapter iii, part A for 'types of innovation')

You can't know the type of innovation till it actually happens in real battlefield. You can't say that you are attempting a sustaining innovation or disruptive (evolutionary or non-evolutionary) innovation. Before it happens, it is just an "innovation".

The very fact that you hear such phrases like "we are aiming disruptive innovation" indicates a serious disconnect with the reality.

That's why I say that it is sad to see that innovation has been reduced to a fad.

Initially all focus is on succeeding, getting it right and not ending up disrupting ourselves. That is the only worry, prayer and endeavour.

Disrupting others is not on the agenda, though a wish. You just innovate, rest happens.

Myth 23

Innovate on the Basis of Customer Surveys

"When the best firms succeeded, they did so because they listened responsively to their customers and invested aggressively. But, paradoxically, when the best firms subsequently failed, it was for the same reasons—they listened responsively to their customers and invested aggressively." – Clayton M. Christensen

This myth has led innumerable companies into tragic deep dungeons of unexplained extinction. The belief behind this myth can be a tragically wrong line to follow, especially blindly and religiously.

Larry Weber had famously said

"Don't listen to you customers. They don't know what they want anyway."

While any new start-up can attribute its heady success to spotting and responding successfully to the customers' unmet and latent needs, it has to compete with the ever evolutionary trajectory of the business model compounded by non-stop changes in technology, processes, environment and customer needs.

Customers often don't know what they want next. They approve of any improvement and make it a huge success only to dump it at the sight of a new, better, interesting and profitable option (which they could never imagine themselves) offered by a new entity on the block.

Henry Ford had said

"If I had asked people what they wanted, they would have said faster horses."

While customers can tell authentically what they like and what they don't (hence accepting or rejecting a product, service or concept), they can rarely tell accurately, and in detail what they would love to have instead. Mob woefully lacks imagination and possibilities, courtesy imagination and creativity are endless!

In the immortal, though intriguing words of Steve Jobs

"A lot of times, people don't know what they want until you show it to them"

"It's not the cusstomer's job to know what they want!"

Satisfied customers fear the loss of their state of satisfaction. This explains why they are averse to the risk of change when asked in a survey or feedback.

That risk is the liability of the entrepreneur and it is the entrepreneur who has to take it.

Marty Cagan had wisely argued – "Customers don't know what they want. It's very hard to envision the solution you want without actually seeing it."

In the words of Jacob Nielsen – " The critivcal failing of user interviews is that you're asking people to either remember past use or speculate on future use of a system."

When a customer says he wants to keep eating cake at tea time, all he is saying is that he is loving it and wants that experience or proposition to continue without jeopardy. What he is certainly not saying is that he will not like to have pizza or something else instead. The customer's approval of the existing product isn't his affidavit that he will not try and love something else if offered and if satisfies him equally or more!

> *Customers don't know what they want. There's plenty of good psychology research that shows that people are not able to accurately predict how they would behave in the future. So asking them, 'Would you buy my product if it had these three features?' or 'How would you react if we changed our product this way?' is a waste of time. They don't know. - Eric Ries*

Investing and betting on the results of customer surveys can be dangerous especially if the customers are yours.

Nobody hated bulk television sets with picture tubes till plasma, lcd or led televisions arrived. Nobody hated the Nokia's and Blackberry's till iPhones and Samsungs arrived. Nobody disliked desktops till laptops came up. Nobody disliked buying from brick and mortar shops till e-Commerce happened.

Examples are littered all around you and so are the corpses of once immortal brands and concepts.

Established companies love customer surveys, especially the ones that endorse them, because they tend to offer solace to the always-fearing-extinction mind of the entrepreneur. Often, that is a false comfort. An entrepreneur is always better off and safer being skeptic than otherwise! Emerging market scenarios don't go back just because you turn your eyes at the assurance of the existing customers.

Customers are good but they have their own interests as their priority. Besides, what they really want what they say they want, are often different...full of hidden and unknown fears, fantasies and prejudices often not backed by commitments.

The fear of cannibalizing sales of existing products is often cited as a reason why established firms delay the introduction of new technologies (even when they are aware and capable of it)...but when established firms wait until a new technology has become commercially mature in its new applications and launch their own version of the technology only in response to an attack on their home markets, the fear of cannibalization can become a self-fulfilling prophecy. – Clayton M. Christensen

Myth 24

Disruptive Innovation needs Technological Breakthrough

Every new technology starts out with the discovery of some phenomenon, even though the utility of that phenomenon usually isn't clear at first. - W. Brian Arthur, *The Nature of Technology*

While a technological breakthrough may cause disruption, it is not that common. In fact, it is unlikely for the reasons discussed below.

Any new technology needs time to get developed, mature, come of age, remove its burrs, connect the dots, discover the missing links and complete the missing parts of the picture before it can be gainfully employed.

Any new technology is not commercially ready on day one. It is not plug-and-play on its birthday; far from it...

And very often, the minds in the incumbent companies and entities are the ones who come up with or notice the developments in technology. They are generally among the first ones to know about the new technology. After all, they are not the market leaders for nothing.

While they are aware of the new technology and have a decent sense of its potential, they and their managements go slow and underplay the newer discoveries because they have to keep the commercial ecosystem and the whole business in mind. There is every chance that they may have over-estimated the potential of the technology or that the new development may be commercially ahead of its time.

There is no scope for any foolish decision which may turn out to be very costly just in the name of being pro-actively "innovative".

No stakeholder wants to disturb the status quo which is yielding great profits. Nobody wants to rock the boat just because of some half baked development on the horizon. In fact, the fear of scoring an own goal is much more than being overtaken by competition. It is a big dilemma.

The obvious path the innovators and their team therefore take is – "wait and watch", not make the first move into the blind dark cave!

This is where the seeds of inevitable blunder get sowed!

The technology anyhow keeps growing rather quite fast, stealthily, step-by-step without taking a breather. The challengers sniff a jackpot and go much beyond the point where the incumbents had deliberately surrendered their leadership.

The incumbents take solace in the argument that the start-ups of new technology will fail. When the start-ups don't fail, the incumbents take solace in the argument that the scale of the challengers and their market-size is too small to be of any concern. By the time, it is realized that the market-size of the challengers is refusing to stop growing, the panic sets in and the very technology that the incumbents had left on the shelves, is scrambled for, implemented and launched, but sadly, as a "me too"! The advantage of being the established dominant player is, thus, dramatically and tragically lost.

While any technological development is potentially a disruptive innovation, it takes so much of time to be one that it is not a surprise for the serious players in the field. It can, therefore, be safely said that it is not the technological breakthrough but the patient process of preparation to exploit its commercial application is what causes the disruption.

Disruptive Innovation, therefore, is not caused by the technological breakthrough, as it may seem at the face of it, but by its innovative use.

The loser incumbents are not undone by the technology but the innovative application of it besides their own complacency and self-denial.

A so-called would-be "disruptive" technology gives a lot of hints, opportunities and warnings before finally getting triggered in the hands of the one who takes the initiative without worrying about the scale, struggle and uncertainty.

In fact, every time the disruptive technology is taken as too small a game to be of interest to any big established player and perhaps the only gate and ray of hope for any challenger.

Ironically, that small breach in the fortress ultimately leads to the crumbling of the empire before the unsuspected handful invaders!

The same process has been repeated time after time since ages!

Howsoever, giants you may become, you have to ensure that you don't lose the spirit and desire to be simultaneous small enough to see the new openings into future.

No big company or entity should forget how it was born, for that's the way for the next one as well. Start-ups are timeless and the only way! Never forget how to be one! Forgetting your past may cost you future!

Besides, it is also to be kept in mind that disruptive innovations are not only about new technologies. Innovative business models, processes, concepts and other non-technical ideas can also be very disruptive!

> **"Disruptive innovations were technologically straightforward. They generally packaged known technologies in a unique architecture and enabled the use of these products in applications previously (technologically or economically) not feasible." – Clayton M. Christensen**

Part D

Building Innovation-Capable Organization

1

Innovation and Organization

"The mere availability of a brilliant new idea for a product or industrial process is not enough. There must also be an ecosystem of firms that are capable of bringing the idea to fruition." – William F. Malony

Discovery of an Innovative idea is just half the battle.

It is a critical half, nevertheless. Still, it is a mere half and a helpless one, without the second half, the implementation.

For both the halves, a team is required, along with its ecosystem and habitat of infrastructure, facilities, processes etc like a seed needing fertile ground, water, air and sunlight!

While an individual can come out with an innovative idea as well execute it himself, yet, at a larger scale innovation assembly line needs the lap and support of a larger team, the organization. Besides, several innovations are beyond the capacity and domain of individuals.

Not only that, what a team can achieve, in terms of magnitude, scale, complexity as well as quality may be simply not possible by a stand-alone warrior.

This is exactly what Scott Berkun is tryng to warn us when he says

> *"Edison didn't work alone – but he did get all the credit. Hundreds of mini inventions paved the way for him – filaments, glass blowing, electricity etc. It is convenient to tell the story of the lone inventor though – Henry Ford, Steve Jobs, etc. Neil Armstrong was the first to walk on the moon, but 500,000 people at NASA helped. Every component in an innovation (eg Television set) represents a long train of inventions."*

While an individual innovator may not need elaborate systems and support to execute his innovations, an organization does need that, lest it should all be chaotic and unsustainable.

> *"Innovation has two significant phases in discovery. One is having that brilliant idea and the other is the delivery of that idea. You can't deliver on an idea unless you have the ability to implement it in a sustainable way...the problem for larger organizations is in the second phase..."* - Anna Kwiatkowska, Former Head of Insight Innovation & Data Science at media giant Sky

In the subsequent pages, some key requirements are being shared which can help design and run a robust system which can inspire, nourish, blossom and sustain innovation, which can help the organization be innovation friendly as well as innovation ready!

2

Securing the Management

The relationship between leaders and teams is one of superior to subordinate. People lower down the career ladder defer to and comply with the wishes of people at more senior levels. In return, leaders protect and reward their people. (Carolyn D., Sherina E. and Michael L., Mckinsey)

The management of any company has the veto power over any matter of it.

Nothing dares to happen if the management doesn't want or back it. And nothing really fails that easily which the management wants to succeed.

Cultivating innovative thinking starts at the top. Leaders can foster a culture of innovation by encouraging creativity and experimenting with new ideas. - Lynne Doughtie

The favourable attitude and alignment of the management towards anything is a deciding factor. Management is the de-facto owner or the one running the show, the one answerable for the outcome (besides everything else) and hence the one empowered to take the red, yellow or green decisions.

Following four scenarios may emerge:

1. Management gives green signal but is not too keen on the thing being done

    - The thing's fate is written in advance. Though nobody will block or interfere, nobody will actively support it as well. There will be enough subtle signals for everyone not to go overboard and "waste" too much energy or resources on the matter. Often, it is management's strategic compulsion to let the thing "happen", a lip service. Innovation taken up by companies this way is highly unfortunate and has little scope or future.

        Everyone down the line senses the unmistakable slackness in approach thereby leading to gradually fading dummy exercises fake-labelled as Innovation.

        Such a situation is highly undesirable as it does more damage than the good it projects to serve.

2. Management doesn't give green signal and neither it is keen on the thing to be done.

    - This is point blank and upfront use of the veto option by the management but is not that bad for the simple reason that it has not been done stealthily and no attempt has been made to double play. Clear signal has been given about the disagreement. There is no ambiguity of management's locus standi and no mixed signals to the team and others.

It is better for managements to block or postpone innovation campaigns in this way instead of covert games. Such a scenario may also arise if a company has some specific or strategic reason behind the same.

3. Management gives the green signal and wants it to be done

   - This is a clear and emphatic positive signal to one and all...an ideal and desirable situation for any endeavour. Everything and anything necessary for the success of the project is reasonably provided. Various forces within the organization are aligned as a vector and things get happen.

   This is an ideal start for the Innovation crusade and a desirable one. Clear signals go to one and all about the intentions of the management, and the direction to be followed.

4. Management doesn't give the green signal but wants it to be done

   - This is a strange but understandable situation. Happens so many times especially in tactical and political situations where it is not feasible or advisable to openly support or allow something to happen which otherwise requires to be done. Doing it covertly becomes a necessity. Subtle but firm signals are there for everyone to support and be with the cause. Resources are ensured though with some obvious constraints.

Such an interesting situation may arise in an Innovation venture when there is no clear backing by the board to the management or, for some tactical reason, a second best scenario for the fate of innovative ventures. Resources are allowed instead of active push; though not ideal, encouraging umbrella cover for the team(s).

It is amply clear from the above brief discussion as to what is best and what is not for the Innovation drives in any organization.

Since innovation is often introduced as an optional add-on to job profiles, functions and roles, it all gets reduced to the supplementary signals from the top as to how to respond.

Signals from the top are infectious and powerful. Even a slightest of positive or negative signal from up there is amplified in the minds of everyone observing.

Nobody's commitment can be more than that of the top management. It is unlikely to be even equal to theirs!

Leadership's attitude and seriousness towards innovation roll-out reveals their true thoughts and intentions regarding the entire campaign. It also reflects the signals from the board of directors or upper hierarchy (stake holders, government, etc).

Not only is the management's commitment expected to be legendary, it has to be inspiring. Its understanding of the need for innovation has to be top rate because that belief only can be the source and force behind the compliance, motivation and application down the line.

The managements not only have to be willing to innovate and let innovate, it has to be visibly and overwhelmingly enthusiastic to throw their weight behind the entire movement. The drive, the thirst has to be omnipresent. They have to be the patrons, the father figure to everyone with them.

Once the management and leadership is secured for the innovation cause, the battle is half won even before the first step is taken.

This single factor is so critical for the entire Innovation ecosystem that it is better to stop altogether and abandon the chase if either management's vision is not clear (because in that case, this needs to be set right first before proceeding even an inch) or if the management is not in a position to lead whole-heartedly.

Company boards and shareholders will tell your CEO that innovation equals risk; what they won't tell her is that innovation never happens unless it comes from the top.
- Skot Carruth

3

Company Wide Innovation Management

Even the strongest dose of the best analgesic on the market won't help mend a broken bone. Likewise, companies can't just import the latest fads in innovation to cure what's ailing them. Instead, they need to consider their existing processes for creating innovations. - (The Innovation Value Chain, hbr.org, Morten T. Hansen & Julian Birkinshaw)

Now, when we are assured of the full backup and leadership of the management, we can focus on the first concrete step towards building an Innovation-friendly and Innovation-ready organization.

As discussed in the 'Myths of Innovation' section, Innovation is not restricted to any one department / function / role but applies to anywhere and everywhere, from sales and marketing to product design, from quality control to purchase, from accounts to maintenance, from logistics to customer care, from planning to HR, everywhere...It is a company-wide concept. So it has to be rooted and practiced company wide.

Therefore, any Innovation system has to engulf and cover all departments / functions / roles and everyone in there. Just like TQM (Total Quality Management), also known as CWQM (Company Wide Quality Management), we have to have TIM /

CWIM (Total Innovation Management) / Company Wide Innovation Management).

It is not to be injected later on or be a mere showcase.
To repeat and recapitulate, main aim of Innovation being
- To remain in the game
- To stay ahead of the competition
- To solve problems
- To keep improving
- (And, if that's on the agenda, go on Disruptive Innovation offensive)

In my studied opinion, Innovation can be assimilated into the company's routine life fastest by making it the newest, grandest and most important part of TQM. In other words, TQM can be made the special purpose vehicle to usher in Innovation.

When I say 'part of TQM' I am talking strictly of TQM/CWQM and not QA or Quality Control or Quality.

The main reason for riding Innovation on the TQM waves is that there is a lot of synergy between the two. In fact, the DNA of both is quite similar.

While TQM, in its most simple definition, is the Assurance-of-Quality-through-Process approach in every function (where Quality is not merely the quality of the product or service but rather, of everything in and about the organization), TIM / CWIM is exploiting those very systems and soft qualities of the organization to further the aims of Innovation mentioned above. It is like building elevated metro rail tracks on the already existing intra-city roads, and thus ensuring a seamless integration of the new transport system in city's lifeline without any disturbance.

Improvements, design breakthroughs, process improvements, problem solving, cost reductions, et al are common to both TQM and Innovation (or TIM).

In fact, Innovation (or TIM) can be considered natural logical higher succession to TQM.

Instead of dismantling all "floors, walls and ceilings" of the organization to "accommodate" and "make way for" Innovation, it must be ensured that Innovation doesn't result in disruption in the organization. It will be a like an attempted suicide. Disruption is an offensive tool aimed at competition, not at one's own temple.

Innovation's challenge is to earn its own respect while to give it a proper introduction, place and chance is ours. It has to run in tandem with the system, get in sync and then accelerate as well as steer it to the next level.

But, what if your organization doesn't already have a TQM or similar framework (like ISO-9000 etc.)? What if there is no formal quality system?

In that case, it is an even bigger opportunity for your organization. A timely wakeup call...

There is a lot to gain and improve in a short time span by initiating the quality journey, never mind the delay!

While Innovation can be practiced and implemented function-wise / depart-wise as standalone cases or from situation to situation and objective to objective, and with handsome rewards, the need for company-wide innovation-compatibility is required for competing with and surviving competition which doesn't have the handicaps your organization has.

4

Innovation Training

"Amazon, a $750 billion company with 566,000 employees doesn't count as a corporation. It acts like a startup" - the world's most innovative companies 2018
(fastcompany.com)

This is not because of its vision statement, fancy marketing or sophisticated software; but because of the Innovation readiness of its human resource.

Now, when the management is leading the crusade and the broad roadmap for "Company Wide Innovation Management" is ready, we have to get the team ready for the journey, the challenge.

Without adequate training about Innovation the entire exercise is likely to fail.

"You have to train people how to be business innovators. If you don't train them, the quality of the ideas that you get in an innovation marketplace is not likely to be high." - Gary Hamel

For this internal as well as external training sources can be tapped. Gradually, the focus has to be on internal trainings for which there have to be regular 'train the trainer' programmes besides continuous efforts to identify, groom and mentor innovation leaders and change agents.

Everyone needs to be trained as everyone is party to the process of Innovation directly or indirectly.

Besides the tools and techniques, the training has to focus on the mindset and the myths, for these constitute the bedrock of Innovation.

Workshops need to be conducted besides considerable hand holding.

5

Innovation Circles

Innovation at Apple has always been a team game. It has always been a case where you have a number of small groups working together. - Jonathan Ive

While idea mining may appear to be an individual game, Innovation process is a team sport.

Coming out with a brilliant idea is only half of an Innovation. Without successful implementation it remains frustratingly incomplete and of little practical use.

While thinking can happen inside every mind strictly individually and separately, it requires critical stimulation by questioning of others.

Besides, ideas from multiple sources can have magical synergy and mutual influence.

So, the best way to make innovation happen on ground is to have Innovation Circles (just like we have Quality Circles).

These are primarily small groups of co-workers (with one or two cross-functional members to bring a different perspective).

These Innovation Circles need to meet periodically and work as per set agenda.

There can be multiple Innovation Circles comprising different mix of same members depending upon the scope and objective.

Having an Innovation coordinator reporting to the plant head (or equivalent) is strongly advised.

6

Innovation Quantification, Targets, Monitoring & Showcasing

There are different ways to do innovation. You can plant a lot of seeds, not be committed to any particular one of them, but just see what grows. And this really isn't how we've approached this. We go mission-first, then focus on the pieces we need and go deep on them and be committed to them. - Mark Zuckerberg

Peter Drucker had said - "You can't manage what you can't measure." It applies perfectly to Innovation as well.

Though Innovation is substantial part an art, its management has to be quite scientific.

The very purpose of the tools, techniques and frames of references for generating innovation ideas indicates that the process of Innovation can be reasonably standardized with the help of some structural frame work.

Innovation can't be in air or in the dark. Measurable parameters of both the Innovation process as well as Innovation projects need to be identified, captured regularly and monitored to form the basis for statistics and dashboard for management.

The work done by the Innovation Circles through the team members need to be adequately recorded / documented and progress monitored. Monitoring has to be non judgemental and inspiringly supportive.

It all has to start with some objective(s). There has to be a pool-in of potential objectives popped up by observations, feedback, problems, non-conformities, competition benchmarks, customer demands, or strategic business goals etc.

If the objectives don't surface out of in natural course of circumstances, observations etc, these need to be forced and taken up artificially in line with strategic short, medium and long term goals of the organization.

The count of Innovative objectives can be one parameter of quantifying innovations. Another way can be function-wise Innovations / Patents / Suggestions etc.

Parameters-based targets for individuals as well as functions can be decided.

All achievements become benchmarks, case studies, motivation points and reference points for all subsequent projects. All innovations of the company must be proudly and properly showcased for there is magic embedded in them which reveals at the right times in future!

All in all, the aim is to have a "wire-mesh" and "point-of-reference" framework for guiding, monitoring, measuring and managing the whole innovation process.

7

Resources

Thanks to shareholder pressures, most CEOs settle for "innovation theater." At many companies, then, the innovation "department" is but a shell with a figurehead. And most CEOs, boards, and investors are content for it to stay that way.
- Skot Carruth

The first resource Innovation will demand the time of the executives for which you are "already paying."

Not only that you will be pulling that much of human resource out of the ongoing projects and tasks. You would have to "spare" them. And that too not for a few hours and only once, (which you could have convinced yourself or your accounts or higher management, pun intended), it is going to be a long term affair, perhaps forever.

Google, I read somewhere, allows its employees as much as 20% of their office time on innovative projects of their own.

Innovation has nothing to do with how many R & D dollars you have. When Apple came up with the Mac, IBM was spending at least 100 times more on R & D. It's not about money. It's about the people you have, how you're led, and how much you get it.
- Steve Jobs

Unless you have the eyes to envision the benefits and unthinkable return on "investment", unless you understand the doors and windows it can open, unless you realize the dangers of not going this way, unless you are sensitive to the reasons why it is important, you will not be able to bear the tremendous loss hinted above (pun again intended)!

This is precisely the reason why I underlined the importance of the management's make or break decisive role in Innovation.

Would like to repeat a few phrases:

"Not only is the management's commitment expected to be legendary, it has to be inspiring."

"The managements not only have to be willing to innovate and let innovate, it has to be visibly and overwhelmingly enthusiastic to throw their weight behind the entire movement."

and

"...it is better to stop altogether and abandon the chase if either management's vision is not clear..."

Unfortunately, majority of the managers or managements are chronically dry and regressive about open culture and open minded measures.

In their "Unleashing the Power of Innovation" report by PWC in 2013, based on the survey of 246 CEOs from across the world, 43% mentioned 'Financial Resources" and 41% "Existing Organizational Culture" as the "constraints" stopping them from being more innovative?

Nobody, including the competition and the economy minds that. While the competition will love to see you go sick and dead, the economy can't help but be on the right side of the survival of the fittest Darwinism.

Besides the time and monetary wages, resources in terms of raw material, logistics support, process spending, training and equipping, facilities, trials etc shall be required and would have to be willingly and merrily extended promptly.

"Innovation requires resources to invest, and you can see many companies pulling back and going into an intense protective mode in a major extended period of financial distress. - Peter Senge"

It has to be understood that this is part of the cost any organization bears to survive and thrive. There is still nothing called a free lunch in this world.

The results and fruits of all this is usually many times more than the cost incurred. Those who are not willing to put in this investment are doomed.

8

Environment, Culture and Ecosystem

Culture translates formal rules into unwritten ones, and it is these unwritten rules that determine behaviours, and guide employees. Any attempt to change culture, without an understanding of the underlying assumptions and repercussions, will fail. – Feng Li

All said and done, crop needs the right soil, water, air and sunlight to grow and remain healthy.

**Innovation and best practices can be sown throughout an organization - but only when they fall on fertile ground.
Marcus Buckingham**

Similarly, the organization has to have the right environment, culture and ecosystem to nurture, back, support and encourage the fruits of Innovation.

"Our goal is to hire talented individuals and then to place them in a culture that allows them to experiment and innovate. In addition, we empower our employees by giving them authority to make decision. This is the key to creating agile and motivated teams." – Steve Basra, Toyota

The air of the organization has to be filled with positivity and not otherwise. You can't fake it. The management or seniors can't fake it. It will be too evident in body language, in the choice of words, the tone, the delay in response, in behaviour, in decisions, in mannerism, in protocols, in ambience, in mood etc.

You can't fake it. It will get revealed magically and understood effortlessly.

If an organization values innovation, you can assume it's safe to speak up with new ideas, leaders will listen, and your voice matters. Adam Grant

How much the management values and cares about your involvement and efforts will show. A frown, a sneer, a quip, a crib etc will expose it all...

In the words of Mike Todasco, Director of Innovation at PayPal,

Innovation stems from curious people asking questions and trying to solve problems, and we need to create that environment where that is encouraged and not discouraged and people are allowed to take calculated chances.

There is an inevitable element of risk in the Innovation zone. And risk is synonymous with potential loss. That loss may be monetary or in any other form...whether that loss is understood as the unavoidable cost of the whole endeavour and encouraged, leave alone frowned at, shall decide whether or not that risk is taken and repeated!

According to Micah Soloman "Employees universally feel safer going with the status quo than attempting innovation, because it's less likely to lead to visible errors (and thus, at many companies, blame). So if you want employees to experiment, they have to know that their innovative efforts will be free from repercussions."

Fear of failure is the enemy of creativity and innovation. Since not all ideas will work, companies have to be bold and encourage a culture where it is ok to fail so long as new ideas are being tried. Even a single good idea that clicks is likely to disproportionately over-compensate for the failure of a dozen more. At times, even a one to thousand success ratio more than justifies all that is endured to innovate! A single idea can change the course of a company, a nation, a planet!

Innovation demands risk-taking - which, in turn, entails redefining failure, stripping away its power to inhibit. - Lynne Doughtie

It is impossible and impractical to attempt calculating the cost of innovation; for there is none. The odds are unfairly tilted towards the benefits, whatever be their odd! A solution to a chronic problem is a blessing even if it were to come in the 101st attempt! In the long run, the cost is likely to be laughably minuscule!

That the risk can always be managed is another point to be kept in mind.

In the words of Jonas Altman (as a part of discussion on Jeff Bezos's "reversible two-way doors" approach) -

"If the decision is not final then it is safe to try. Adopting this type of mental model is necessary, as both experimentation and failure are part and parcel of playing the innovation game today."

Any organization can trick, coerce or mislead the human resource to Innovation once. However, it is their experience, word of mouth verdict and resulting attitude that will decide whether follow-up Innovations will happen or not.

Anna Kwiatkowska goes to the extent of prompting the culture of Intrapreneurship by defining it as "the mentality (of the employee) of acting like an entrepreneur in a larger, mature organization…" It is worth pondering whether this is possible in adverse culture and environment?

Dictionary defines 'Culture' as "the ideas, customs, attitudes and social behaviour of a particular or society."

The words "Customs", "Attitudes" and "Behaviour" hold the key to the Innovation lock of any organization.

And most of this trickles top-down!

And is quite fragile!

Tolerance, risk acknowledgement and patience for results are holy touchstones of Innovation-friendly culture and environment.

Dr.Phil Mcgraw's classic quote that "we teach people how to treat us!" wonderfully sums up what the organizations say or imply (directly or otherwise, subtly or otherwise) and what lesson it carries for its human resource.

According to Faye Holland, "The thing that always works is taking people out of their normal environment and welcoming a variety of perspectives (read cross-functional team)."

> **"Innovation magic occurs at the very moment that each member of an innovation team believes that they have full freedom to add their best ideas, while top management believes that it is in complete control; both, at the same time." - Bill Fischer**

9

Motivation, Appreciation, Rewards and Recognition

Haier, a Chinese multinational consumer electronics and home appliances company, started naming its products and services after the person who proposed the idea that ended up being developed. – (Sophia Hübner, 7 Habits of Highly Innovative Companies, www.itonics.de)

What gets rewarded gets repeated, they say!

Innovation is a fit outcome to be repeated and hence the need for rewarding the performances.

As aptly put by John Adair in his classic "Effective Innovation",

"Don't blame your staff for lack of interest or lack of new ideas. There are no unmotivated or uninnovative people at work, only poor managers."

In the words of Julian Birkinshaw,

"Employees face capacity, time and motivation issues around their participation. There is often a lack of follow-through in well-intentioned schemes...a disconnect between the priorities of those at the top and the efforts of those lower down in the organization."

While sincere and qualified appreciation is the most economical and instantaneous reward and that too with astonishingly high "return on investment", motivation and recognition in any form is a magic manure for the Innovation fields.

Innovation epitomizes the highest of the human needs...self realization, and it represents the highest of corporate excellence...extended immortality!

External Challenges

Macro Economic Environment

Which Government / Political Stability

Credit Availability

Labour Laws / Issues

Market Conditions

Competition Scenario

Technology Scenario

Regulatory / Policy Environment

Bureaucracy

External Costing Pressures / Raw Material

Internal Challenges

Problem Solving

Handling Sticky Situations

Manpower / HR

Quality Management

Processes, Productivity challenges

Costing pressures

Product / Service Design

Finance / Credit / Money Circulation

Sales, Marketing, Distribution, Logistics

Unforeseen Problems

Business Decisions

Epilogue

Facts and Figures

- 41% of respondent corporate strategy executives admitted that their organizations were at extreme risk of disruption - "State of Innovation" report by CB Insights

- 78% of Innovation portfolios allocated to sustaining innovation instead of disruptive risks - "State of Innovation" report by CB Insights

-

Wise Words from Innovation Masters

As companies get bigger, they typically get dumber. They are less innovative and more bureaucratic. – Jonas Altman

***

Innovation is difficult because it demands new work. But, at a more basic level, it's difficult because it requires an admission that the way you've done things is no longer viable. - Mike Shipulski

***

"Don't worry about people stealing your ideas. If your ideas are any good, you'll have to ram them down people's throats." - Howard Aiken

***

Without tradition, art is a flock of sheep without a shepherd. Without innovation, it is a corpse. - Winston Churchill

***

"Despite the vast potential of technology to spur economic growth, global surveys suggest that developing-country businesses do far less than expected to adopt advanced-country techniques to upgrade products, technologies, and business processes...improving management practices and undertaking the basic research and development that is necessary to adapt technologies to the local context." - William F Maloney (Business-Standard.com)

***

The regulatory systems in place dis-incentivize innovation. It's intense to fight the red tape. - Travis Kalanick

***

I said from the very beginning, 'Yahoo should position itself as a technology innovation company, not as a media company.' - Masayoshi Son

***

The QWERTY keyboard was invented to stop typewriters sticking, but the design has stuck. Better designs might follow, but to gain acceptance, they must improve on that dominant idea by a sufficient margin to justify the cost of the switch (eg relearning how to type). – Scott Berkun

***

Make sure your appetite for innovation matches your overall targets for growth. – PWC

***

Someone comes to you with their new idea, and you first wait 24 seconds before responding. If you master that, try and wait 24 minutes before responding next time, with the ultimate aim being to wait a full 24 hours before you respond with feedback on the new idea. – Anthony Tjan

***

Failing a lot is ok, because one or two "winners" can cover for hundreds and thousands of failures. – Jeff Bezos

***

"To live a creative life, we must lose our fear of being wrong."
- Joseph Chilton Pierce

***

All business boils down to two things: Marketing and Innovation. Businesses that fail to innovate fail to thrive, and eventually, fail to exist. At best, they become obsolete. – Peter Drucker

***

The mentality that got you to the current level of success will not get you to the next level. Even at the individual level, innovation, which is simply a new idea put into action, is critical for growth and fulfilment. – Ashley Good

***

The best ideas come as jokes. Make your thinking as funny as possible – David Ogilvy

***

www.ingramcontent.com/pod-product-compliance
Lightning Source LLC
Chambersburg PA
CBHW030618220526
45463CB00004B/1342